Explorations in College Algebra and Trigonometry Using the TI 83/83 Plus/84 Plus/86

Third Edition

Deborah Jolly Cochener
Austin Peay State University

Bonnie MacLean Hodge
Austin Peay State University

THOMSON
BROOKS/COLE

Australia • Canada • Mexico • Singapore • Spain • United Kingdom • United States

Printed in the United States of America
1 2 3 4 5 6 7 08 07 06 05 04

Printer: Thomson/West

ISBN: 0-534-42262-4

Thomson Brooks/Cole
10 Davis Drive
Belmont, CA 94002-3098
USA

Asia
Thomson Learning
5 Shenton Way #01-01
UIC Building
Singapore 068808

Australia/New Zealand
Thomson Learning
102 Dodds Street
Southbank, Victoria 3006
Australia

Canada
Nelson
1120 Birchmount Road
Toronto, Ontario M1K 5G4
Canada

Europe/Middle East/South Africa
Thomson Learning
High Holborn House
50/51 Bedford Row
London WC1R 4LR
United Kingdom

Latin America
Thomson Learning
Seneca, 53
Colonia Polanco
11560 Mexico D.F.
Mexico

Spain/Portugal
Paraninfo
Calle/Magallanes, 25
28015 Madrid, Spain

TABLE OF CONTENTS

Graphing and Exploring Equations in Two Variables

Trigonometric Functions

Introduction of Keys

Unit Title	TI-83/84 Series Keys	TI-86 Keys
#1: Getting Acquainted With Your Calculator, p.1	ON/OFF 2nd ▲ ▼ (to darken/lighten) MODE ENTER CLEAR (-) CATALOG MATH ▸NUM 1: abs(( ) parentheses $\sqrt{}$ x^2 ∧ QUIT INS DEL π ENTRY ANS	ON/OFF 2nd ▲ ▼ (to darken/lighten) MODE ENTER CLEAR (-) CATALOG CUSTOM abs ( ) parentheses $\sqrt{}$ x^2 ∧ QUIT/EXIT INS DEL π ENTRY ANS
#2: Fractions and the Order of Operations, p. 11	MATH 1:▸Frac	CUSTOM▸Frac
#3: Evaluating Through TABLES and the STOre Feature, p.19	STO▸ X,T,θ,n ALPHA : Y= TABLE	STO▸ x-VAR ALPHA : y(x)= TABLE (TI-86)
#4: Rational Exponents and Radicals, p. 29	MATH 4: $\sqrt[3]{}$ (5: $\sqrt[x]{}$	CUSTOM $\sqrt[x]{}$
#5: TI-83/84 Series and TI-/86: Complex Numbers, p. 33	i MATH ▸ CPX MODE Real▸ a+bi	(a,b) = a +bi CPLX
#6: Graphical Solutions: Linear Equations, p.37	Y= WINDOW TRACE GRAPH CALC 5:intersect QUIT ZOOM 6:ZStandard	GRAPH y(x)= RANGE TRACE GRAPH MATH/ISECT EXIT MORE GRAPH/ZOOM ZSTD

Unit Title	TI-83/84 Series Keys	TI-86 Keys
#7: Graphical Solutions: Absolute Value Eq., p.45	No New Keys	No New Keys
#8: Graphical Solutions: Quadratic and Higher Degree Equations, p.48	CALC 2:zero	GRAPH/MATH ROOT
#9: Graphical Solutions: Radical Equations, p.55	No new keys	No new keys
#10: Graphical Solutions: Linear Inequalities, p.59	Graph Style Icon	Graph Style Icon
#11: Graphical Solutions: Absolute Val. Ineq., p.65	No new keys	No new keys
#12: Graphical Solutions: Quadratic Inequalities, p.71	No new keys	No new keys
#13: Graphical Solutions: Rational Inequalities, p. 77	MODE Connected/Dot	GRAPH FORMAT DrawDot
#14: Graphing Basics, p.81	GRAPH WINDOW ZOOM 6:ZStandard CALC 1:Value	GRAPH WIND GRAPH/ZOOM ZSTD GRAPH EVAL
#15: Calculator Viewing Windows, p.87	MODE FORMAT WINDOW ZOOM 1:ZBox 2:Zoom In 3:Zoom Out 4:ZDecimal 5:ZSquare 6:ZStandard 8:ZInteger Graph Style Icon	MODE FORMAT WIND GRAPH/ZOOM BOX ZIN ZOUT ZDECM ZSQR ZSTD ZINT Graph Style Icon GRAPH/SELCT
#16: Functions, p.97	DRAW 1:ClrDraw 8:DrawInv VARS	GRAPH/MORE DRAW CLDRW DrInv

Unit Title	TI-83/84 Series Keys	TI-86 Keys
#17: Translating, Stretching, and Shrinking Graphs of Functions, p. 103	CALC 3:minimum 4:maximum	Graph/MATH FMIN FMAX
#18: Symmetry of Functions, p. 109	No new keys	No new keys
#19: Piecewise Functions, p. 113	TEST	TEST
#20: Exponential and Logarithmic Functions, p. 125	e^x LOG LN	e^x LOG LN
#21: Curve of Best Fit, p. 125	STAT 1:Edit 4:ClrList STAT PLOT 1:Plot 1 MEM 4:ClrAllLists ZOOM 9:ZoomStat STAT/CALC 4:LinReg(ax+b) 5:QuadReg 6:CubicReg VARS 5:Statistics/EQ 1:RegEQ	STAT EDIT CLRxy DRAW SCAT ZOOM ZDATA STAT/CALC LINR P2Reg P3Reg STAT/ VARS RegEq
#22: Matrices, p. 137	MATRX EDIT MATH 1:det(5:identity(A: ref(B: rref(C:rowSwap(D:row + (E:* row (F: *row + (x^{-1}	MATRX EDIT MATH det OPS Ident ref rref rSwap rAdd multR mRAdd x^{-1}
#23: Trigonometric Functions: Amplitude, Phase Shifts, and Translations, p. 151	ZOOM 7:ZTrig MODE Degree	GRAPH/ZOOM ZTRIG MODE Degree
#24: Graphical Explorations: Trigonometric Identities, p. 161	No new keys	No new keys

Unit Title	TI-83/84 Series Keys	TI-86 Keys
#25: Graphical Solutions: Trigonometric Equations, p. 165	No new keys	No new keys
#26: Polar Graphing, p. 169	MODE Pol	MODE Pol PolarC
#27: Parametric Graphs of Conics, p. 175	MODE Par	MODE Param
#28: Roots of Complex Numbers, p. 181	No new keys	No new keys

PREFACE

This unique workbook/text provides the student the opportunity for guided exploration of topics in college algebra using Texas Instruments graphing calculators. Keystroking guidance and correlating concept charts are provided for the TI-83/84 series and TI-86 calculators. This enables the instructor to use the text in a classroom that requires any of the above listed calculators, as well as in the classroom where calculators are mixed. The text is intended for use as a supplemental text to a CORE classroom text, and is therefore arranged by topics. This enables the instructor to assign the appropriate *Explorations Unit(s)* that correlate(s) with the topic under discussion within the classroom. **The text is not meant to be worked in sequential order. Each unit has one or more prerequisite units. *Only* the prerequisite units are required for student success in working the assigned unit.** This allows the use of this ancillary text with virtually *any* core course textbook. Charts that correlate the *Exploration Units* with textbook sections are provided, upon request, from Thomson Learning.

Changes in the Third Edition

▸ The TI-83/84 series are the base calculators used in the text, with changes noted for the TI-86.

▸ Units have been restructured to address topics numerically, graphically, and algebraically.

▸ Increased emphasis has been placed on the interpretation of both graphical and table displays.

▸ The unit previously entitled *Line of Best Fit* has been expanded to examine the *Curve of Best Fit*.

▸ Examples have been restructured in tables with keystroking information provided in a manner that is more easily understood by the student.

▸ Units have been combined to better reflect current pedagogy. For example, the second edition explored parabolas that were functions as well as those that were not. A separate unit addressed translating and stretching graphs. In this edition the unit addressing parabolas that were not functions was deleted and the other two units were combined.

Features

*Each unit provides guided exploration of a topic. **Units are not meant to be done in numerical order, but rather according to concept.** Prerequisite units are listed at the beginning of each unit and appear in the Table of Contents.

*The text requires *no* instruction - but serves as a workbook for the student. Answers to questions/exercises within a unit appear at the end of the unit.

*Many units can be used in place of classroom instruction.

*A key correlation chart that shows the units in which keys are introduced is provided.

*The workbook may be used over a period of more than one semester/quarter as the student progresses through his/her mathematics sequence.

*As an institution changes textbooks, this supplement need not be changed.

*Units are written in a manner that enables the student with little or no algebraic experience to read and explore independently.

*The units provide springboards for both classroom discussion and further investigation either individually or as a class.

Integration of technology into the mathematics classroom has grown significantly with the introduction of the graphing calculator. These calculators have brought relatively inexpensive technology into the students' hands. However, instructors are now faced with the problem of integrating the technology without sacrificing course content. These activities will enable the students to develop algorithms typically found in college algebra courses, improving both their understanding and retention of the material.

The workbook was written with the belief that students who are active contributors in the classroom increase their own understanding and their long term retention of material. Of primary importance, however, is the fact that the units have provided the means and the opportunity for students to create their own mathematics.

Organization

The workbook is divided into six sections.

UNITS #1-5: These units introduce the student to the calculator and its capabilities in performing computational tasks.
UNITS #6-9: These units introduce and utilize the INTERSECT and ROOT/ZERO features to graphically solve linear, absolute value, quadratic, higher degree, and radical equations. The units can be worked before the text formally introduces graphing. However, instructors may wish to assign Units 14 and 15 prior to Unit 6.
UNITS #10-13: These units present the interpretation of tables and graphs to solve linear, absolute value, quadratic, and rational inequalities.
UNITS #14-22: These units introduce the student to graphing as a formal process, introducing the students to ZOOM options and ways to obtain good graphical displays. Emphasis in this section is on functions. The unit on Matrices is also included in this section.
UNITS #23-28: These units explore trigonometric functions.

ACKNOWLEDGEMENTS

We are grateful and appreciative for the input of time and expertise by those who reviewed this workbook:

Tammy Higson, Hillsborough Community College (CAT)
Diane Koenig, Rock Valley College (CAT)
Arnavaz Taraporevala, CUNY-New York City College of Technology (Precalc)
Charles Laws, Cleveland State Community College (Precalc)

We appreciate the guidance and support of Rachael Sturgeon, our editor at Brooks/Cole publishing. We would also like to thank our husbands, David and Blaine, and our children, Sherah, Blaire, and Trey, for their support and love. They have always believed in us and it is they who made the dream become reality. Our families' contributions to the editing, reviewing, wording, and the testing of material has been invaluable.

UNIT 1
GETTING ACQUAINTED WITH YOUR CALCULATOR

*This unit is a prerequisite for all other units in the text. Answers to all units appear at the end of the unit.

Features may be accessed in three places: on the keypad, under a menu, or in the cataolog.

TI-84 If using the TI-84, go to the TI-84 guidelines which follow this unit (pg.7).

TI-86 IF USING THE TI-86 GO TO THE TI-86 GUIDELINES WHICH FOLLOW THIS UNIT (PG.9).

Touring the TI-83/83plus

Take a few minutes to study the TI-83graphing calculator. The keys are color-coded and positioned in a way that is user friendly. Notice there are dark blue, black, and gray keys, along with a single yellow key and a single green key.

<u>Dark blue keys</u>: On the right and across the top of the calculator are the dark blue keys. At the top right are four directional cursor keys. These may be used to move the cursor on the screen in the direction of the arrow printed on the key. The four arithmetic operation symbols are also in dark blue. Notice the key marked **ENTER**. This is used to activate entered commands, thus there is no key on the face of the calculator with the equal sign printed on it. Below the screen are five keys labeled **[Y=], [WINDOW], [ZOOM], [TRACE],** and **[GRAPH]**. These keys are positioned together below the screen because they are used for graphing functions. The TI-83plus has an additional blue key labeled **APPS**. It is used to access the finance menu and extended capabilities of the calculator.

<u>Black keys</u>: The majority of the keys on the calculator are black. Notice the **[X,T,θ,n]** key in the second row and second column. It will be used frequently in algebra to enter the variable **X**. The **ON** key is the black key located in the bottom left position.

<u>Gray keys</u>: The twelve gray keys that are clustered at the bottom center are used to enter digits, a decimal point, or a negative sign.

<u>Yellow key</u>: The yellow key is the **2nd** key located in the upper-left position. To access a symbol in yellow (printed above any of the keys) first press the yellow **[2nd]** key, and then the key BELOW the symbol (function) to be accessed. For example, to turn the calculator **OFF** notice that the word **OFF** is printed in yellow above the **ON** key. Therefore, press the keys **[2nd] [ON]** to turn the calculator off. These keystrokes are done *sequentially* - not simultaneously. Throughout this book the following symbolism will be used: symbols that appear on the key will be denoted in brackets, **[ ]**, whereas symbols written above the key will be denoted in **< >** and menu options will be in parentheses, **( )**. Thus, the previous command for turning the calculator off would appear as **[2nd] <OFF>**. The symbols **[ ]** , **< >** , or **( )** will cue you <u>where</u> to look for a command - either printed on a key, above it, or as a menu option.

<u>Green key</u>: Alphabet letters (printed in green above some of the keys), or any other symbols or words printed in green above a key are accessed by first pressing the green **ALPHA** key and then the key below the desired letter/symbol/word. The keystrokes are sequential.

<u>Catalog feature</u>: Press **[2nd] <CATALOG>** to display an alphabetical list of available calculator operations. Use the **[▲]** and **[▼]** cursor keys to scroll through this list. Operations may be accessed by placing the pointer adjacent to the operation and pressing **[ENTER].**

*Note: The TI-83/83plus has an **A**utomatic **P**ower **D**own (**APD**) feature which turns the calculator off when no keys have been pressed for several minutes. When this happens, press **[ON]** to access the last screen used.*

Let's Get Started!

Turn the calculator on by pressing **[ON]**. If the display is not clear, press **[2nd] [▲]** to darken the screen, or **[2nd] [▼]** to lighten the screen. Notice that when the **[2nd]** key is pressed, an arrow pointing up appears on the blinking cursor.

To ensure the calculator is in the desired mode, press **[MODE]**. All of the options on the far left should be highlighted. If not, use the **[▼]** to place the blinking cursor on the appropriate entry and press **[ENTER]**. Exit MODE by pressing **[CLEAR]**. This accesses the home screen (where expressions are entered). Press **[CLEAR]** until the screen is cleared except for the blinking cursor in the top left corner.

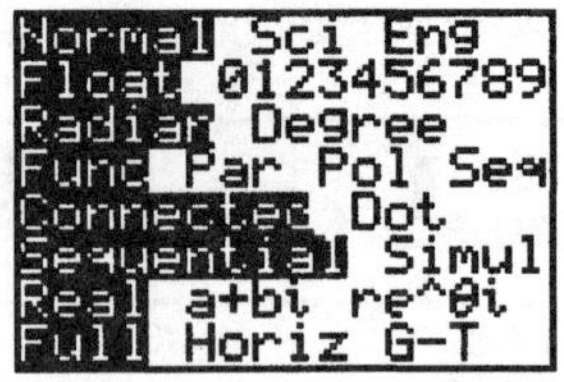

Integer Operations

When entering integers on the calculator, differentiation must be made between a subtraction sign and a negative sign. Notice that the subtraction sign appears on the right side of the calculator, with the other arithmetic operations. The negative sign appears to the left of the **[ENTER]** key and is labeled **(-)**.

Example 1: Simplify: -8 - 2

Expression	Verbally	Graphing Calculator	
-8 - 2	Negative eight minus two.	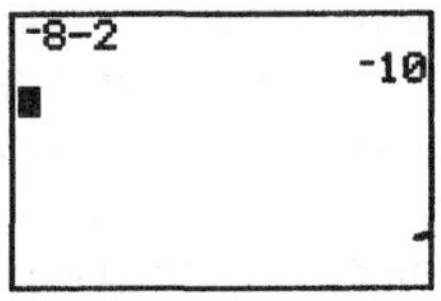 Observe the difference in <u>size</u> and <u>position</u> of the negative sign and the subtraction sign.	**Keystrokes:** **[(-)] [8] [-] [2] [ENTER]**

TI-86 TI-86 USERS TURN TO "ABSOLUTE VALUE" IN THE GUIDELINES (PG.10).

Absolute Value

Example 2: Simplify: |-3 - 2|

Expression	Verbally	Graphing Calculator	
|-3 - 2|	The absolute value of the quantity negative three minus two.	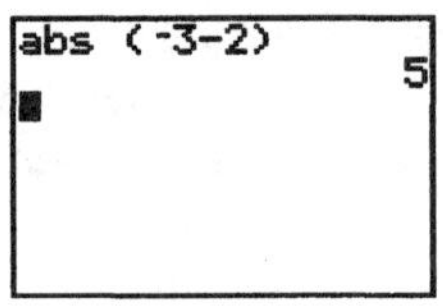	**Keystrokes:** **[2nd] [MATH] [▸](NUM) [1] (1:abs() [(-)] [3] [-] [2] [ )] [ENTER]** Absolute value may also be accessed through the CATALOG feature (located above the 0 key). Press **[2nd]<CATALOG>**. Because the cursor (▸) is pointing to **abs(** , press **[ENTER]** to access it.

Square Roots

Example 3: Simplify: $\sqrt{9+16}$

Expression	Verbally	Graphing Calculator	
$\sqrt{9+16}$	The square root of the quantity, nine plus 16.	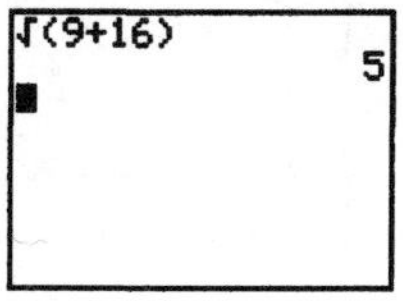	**Keystrokes:** **[2nd] < √ > [9] [+] [1] [6] [ )] [ENTER]** *Note: Square root is located above the* ***[x²]*** *key.*

TI-86 users must enter both parentheses under the square root symbol.

Raising to Powers

A number may be squared (raised to the second power) by either pressing the **[x²]** key after entering the number, or by using the caret **[^]** key and entering the desired exponent. Care must be taken to clearly identify the base.

Example 4: Simplify: a. 4^2 b. -4^2 c. $(-4)^2$

Expression	Verbally	Graphing Calculator	
a. 4^2	Four squared.	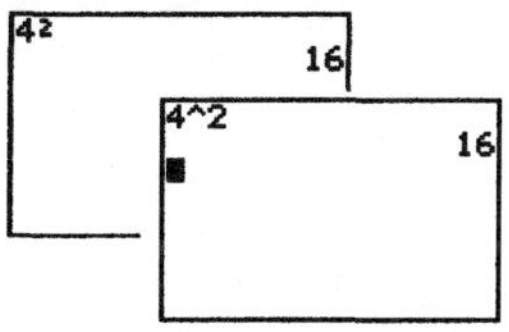	**Keystrokes:** **[4] [x²] [ENTER]** **OR** **[4] [^] [2] [ENTER]**
b. -4^2	The opposite of 4 squared.	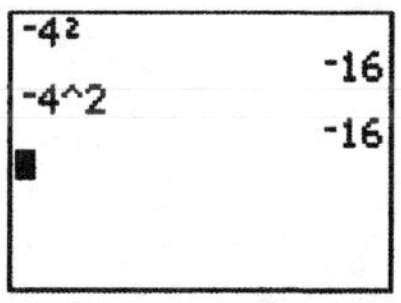	**Keystrokes:** **[(-)] [4] [x²] [ENTER]** **OR** **[(-)] [4] [^] [2] ENTER]**
c. $(-4)^2$	The quantity, negative four, squared.	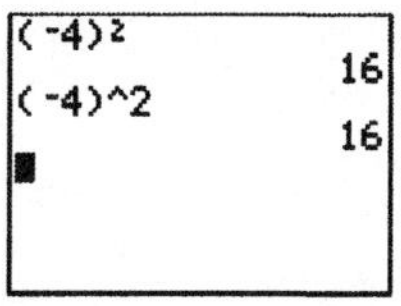	**Keystrokes:** **[( ] [(-)] [4] [ )] [x²] [ENTER]** **OR** **[( ] [(-)] [4] [ )] [^] [2] [ENTER]**

To raise to the third power (or higher), use the caret key.

Example 5 : Simplify: $4(3)^5$

Expression	Verbally	Graphing Calculator	
$4(3)^5$	Four times three to the fifth power.	4(3)^5 972	**Keystrokes:** **[4] [( ] [3] [ )] [^] [5]**

EXERCISE SET

Simplify the following expressions on the calculator. Use the box provided below each problem to **RECORD THE CALCULATOR SCREEN** **LINE BY LINE EXACTLY** as it appears.

1. $|4^5 - (-6^2)|$

2. $|-4.7 - 5.28| - 18.3$

3. $-|-2^4 - 7|$

4. $\sqrt{5^3 - 10^2}$

5. $-\sqrt{169 - 25}$

6. $\sqrt{(-5)^2 - 4^2} + 16$

7. $(15 - 2)^3$

8. $|7 - 11| - \sqrt{64}$

✍ 9. Explain the importance of using parentheses when entering expressions containing absolute value, roots, and exponents.

✍ 10. Explain why entering - 8, negative eight, as "minus 8" (using the subtraction key) is incorrect. Explain the result this incorrect keystroking produces.

✍ 11. a. Explain the meaning of the syntax error message.

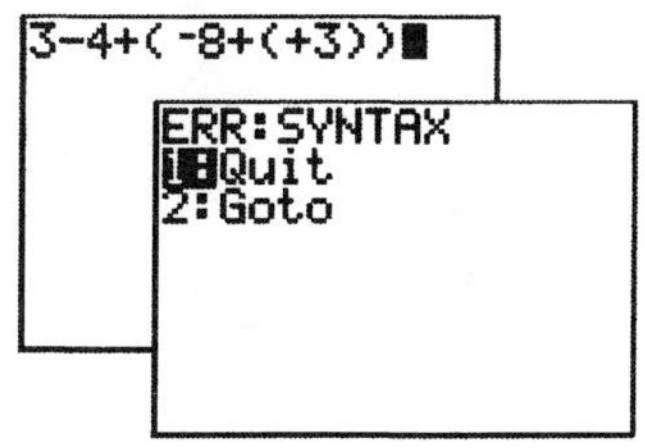

b. What was the error with the arithmetic entry?

Shortcut Keys

Below are descriptions of some keys that may be helpful in the efficient use of the calculator. You may want to reference this material in the future.

QUIT (Located above the key next to the 2nd key - EXIT on the TI-86; MODE on the TI-83/84 series)

To return to the home screen press **[2nd]** **<QUIT>**. This is helpful when you are stuck on a screen and pressing **[CLEAR]** does not return the home screen.

TI-86	THE **[EXIT]** KEY ON THE TI-86 WILL REMOVE MENUS FROM THE SCREEN.

INSERT (Located to the right of the 2nd key, above the DEL - delete - key.)

This key is helpful when data is entered incorrectly - particularly in expressions that are lengthy. When using the insert key, first place the cursor in the position in which the inserted digit/symbol should appear, press **[2nd]** **<INS>** (the character under the cursor will blink) and then the desired digit/symbol(s) to be inserted. The calculator will insert as many characters as desired as long as a cursor key is not pressed.

DELETE (Located two keys to the right of the 2nd key.)

This key is helpful when an incorrect key is mistakenly pressed. Place the cursor over the character to be deleted and press **[DEL]**.

PI (π) (Located above the caret, ∧, key on the right side of the calculator.)

When evaluating formulas requiring the use of π , press **[2nd]** **< π >**. Although only nine decimal places are displayed, the calculator will evaluate the expression using an eleven decimal place approximation for π .

TI-86	THE TI-86 DISPLAYS AN ELEVEN DECIMAL PLACE APPROXIMATION FOR π BUT USES A THIRTEEN DECIMAL PLACE APPROXIMATION FOR COMPUTATION PURPOSES.

ANSWER (ANS is located above the key used to enter a negative sign.)

The calculator will recall the answer from a previous computation. Access this function by pressing **[2nd] <ANS>**. It will also be activated if you press an operation key (+, -, X, ÷) before entering a number.

ENTRY (Located above the ENTER key.)

Pressing **[2nd] <ENTRY>** accesses the ability of the calculator to recall the expression previously entered. Pressing **[2nd] <ENTRY>** repeatedly, performs "deep recall" by scrolling up the screen.

Pressing **[2nd] <◂>** moves the cursor to the beginning of the entry, whereas pressing **[2nd]<▸>** moves the cursor to the end of the entry.

Solutions:

1. 1060 **2.** - 8.32 **3.** - 23 **4.** 5 **5.** - 12 **6.** 19 **7.** 2197

8. - 4 **9.** Absolute value and root symbols are grouping symbols, therefore, parentheses are required to designate the grouped quantity. An exponent only applies to a single number/character unless grouping is used.

10. "Minus 8" indicates the operation of subtracting 8 from another quantity, whereas "negative 8" specifies the opposite of the number 8. Therefore, when a minus sign is used as the initial entry the calculator will retrieve the previous answer (ANS) for the 8 to be subtracted from.

11. a. A syntax error message indicates you have used a symbol incorrectly. **b.** The plus sign in front of the 3 should not have been entered because"plus" indicates the operation of addition and never a positive sign on the calculator.

TI-84 GUIDELINES UNIT 1

Touring the TI-84

Take a few minutes to study the TI-84 graphing calculator. The keys are color-coded and positioned in a way that is user friendly.

<u>**Light gray keys:**</u> On the right side of the calculator are the dark blue keys. At the top are four directional cursor keys. These may be used to move the cursor on the screen in the direction of the arrow printed on the key. The four operation symbols (addition, subtraction, multiplication, and division) are also in dark blue. Notice the key marked **ENTER**. This will be used to activate commands that have been entered; thus there is no key on the face of the calculator with the equal sign printed on it.
Below the screen are five keys labeled **[Y=], [WINDOW], [ZOOM], [TRACE],** and **[GRAPH]**. These keys are positioned together below the screen because they are used for graphing functions.

<u>**White keys:**</u> The twelve gray keys that are clustered at the bottom center are used to enter digits, a decimal point, or a negative sign.

<u>**Black keys:**</u> The remaining the keys on the calculator are black. Notice the **[X,T,θ, *n*]** key in the second row and second column. It will be used frequently in algebra to enter the variable **X**. The **ON** key is the black key located in the bottom left position. The key labeled in purple as APPS is used to access the finance menu and extended capabilities of the calculator.

<u>**Light blue key:**</u> The only light blue key is the **2nd** key located in the upper-left position.

Above most of the keys are words and/or symbols printed in either light blue or green. To access a symbol in light blue (printed above any of the keys) first press the light blue **[2nd]** key, and then the key BELOW the symbol (function) to be accessed. For example, to turn the calculator **OFF** notice that the word **OFF** is printed in light blue above the **ON** key. Therefore, press the keys **[2nd] [ON]** to turn the calculator off. These keystrokes are done *sequentially* - not simultaneously. Throughout this book the following symbolism will be used: symbols that appear on the key will be denoted in brackets, **[ ]**, whereas symbols written above the key will be denoted in **< >** and menu options will be in parentheses, **( )**. Thus, the previous command for turning the calculator off would appear as **[2nd] <OFF>**. The symbols **[ ] , < > ,** or **( )** will cue you <u>where</u> to look for a command - either printed on a key, above it, or as a menu option.

<u>**Green key**</u>: Alphabet letters and other symbols printed in white above some of the keys are accessed by first pressing the gray **ALPHA** key and then the key *below* the desired letter or symbol. Again, the keystrokes are sequential.

<u>**Catalog feature**</u>: Press **[2nd] <CATALOG>** to display an alphabetical list of available calculator operations. Use the **[▲]** and **[▼]** cursor keys to scroll through this list. Operations may be accessed by placing the pointer adjacent to the desired feature and pressing **[ENTER]**.

Note: The TI-84 has an ***Automatic Power Down (APD)*** *feature which turns the calculator off when no keys have been pressed for several minutes. When this happens, press* ***[ON]*** *to access the last screen used.*

Let's Get Started!

Turn the calculator on by pressing **[ON]**. If the display is not clear, press **[2nd] [▲]** to darken the screen, or **[2nd] [▼]** to lighten the screen. Notice that when the **[2nd]** key is pressed, an arrow pointing up appears on the blinking cursor.

To ensure the calculator is in the desired mode, press **[MODE]**. All of the options on the far left should be highlighted. If not, use the **[▼]** to place the blinking cursor on the appropriate entry and press **[ENTER]**. Exit MODE by pressing **[CLEAR]**. This accesses the home screen which is where expressions are entered. Press **[CLEAR]** until the screen is cleared except for the blinking cursor in the top left corner.

☞ Return to the core unit section entitled "Integer Operations" (pg.2) and complete the unit.

TI-86 GUIDELINES UNIT 1

Touring the TI-86

Take a few minutes to study the TI-86 graphing calculator. The keys are color-coded and positioned in a way that is user friendly.

<u>**Gray keys**</u>: The twelve gray keys that are clustered at the bottom center are used to enter digits, a decimal point, and a negative sign. At the top right are four directional cursor keys. These may be used to move the cursor on the screen in the direction of the arrow.

<u>**Black keys**</u>: The majority of the keys on the calculator are black. The four operation symbols (division, multiplication, subtraction and addition) are located in the column on the far right. Notice the key marked **ENTER** at the bottom right position. This will be used to activate commands that have been entered. The equal sign on the face of the calculator is NOT used for computation. The key marked **x-VAR** in the second row, second column will be used frequently to enter the variable **x**. The **ON** key is in the bottom left position.

Below the screen are five black keys labeled **F1**, **F2**, **F3**, **F4** and **F5**. These are menu keys and will be addressed as they are needed.

<u>**Yellow-orange key**</u>: This key is located in the top-left position and is labeled **2nd**.

<u>**Blue key**</u>: This key is labeled **ALPHA** and is located at the top left of the key pad.

Above most of the keys are words and/or symbols printed in either yellow-orange or blue. To access a symbol in yellow-orange (printed above any of the keys), first press the yellow-orange **[2nd]** key, and then the key BELOW the symbol (function) you wish to access. For example, to turn the calculator **OFF** notice that the word **OFF** is printed in yellow-orange above the **ON** key. Therefore, press the **2nd** key and the **ON** key to turn the calculator off. These keystrokes are done <u>*sequentially*</u> - not simultaneously. Throughout this book the following symbolism will be used: symbols that appear on the key will be denoted in brackets, **[ ]**, whereas symbols written above the key will be denoted in **< >**. A function that is accessed from a menu will be written in parentheses, **( )**. Thus, the previous command for turning the calculator off would appear as **[2nd] <OFF>**. The symbols **[ ]**, **< >**, or **( )** will cue you <u>where</u> to look for a command - either printed on a key, above it, or as a menu option. When a menu key is indicated, the current function of the key will follow in parentheses to correspond to the display at the bottom of the screen. For example, press **[GRAPH]** to display the graph menu. The notation **[F2] (WIND)** denotes access to the WINDOW submenu. Users should be aware that the menu denoted as WIND on the TI-86 corresponds to the WINDOW menu referred to for the TI-83/84 series. To remove the menu at the bottom of the screen, press **[EXIT]**.

To access a symbol printed in blue above some of the keys, first press the blue **ALPHA** key and then the key *below* the desired letter or symbol. Again, the keystrokes are sequential.
*NOTE: The **A**utomatic **P**ower **D**own (APD) feature turns the calculator off when no keys have been pressed for several minutes. When this happens, press **[ON]** to access the last screen used.*

Let's Get Started!

Turn the calculator on by pressing **[ON]**. If the display is not clear, press **[2nd]** and hold down **[▲]** to darken the screen or **[2nd] [▼]** to lighten. Notice that when the **[2nd]** key is pressed an arrow pointing up appears on the blinking cursor.

To ensure the calculator is in the desired mode, press **[2nd] <MODE>**. All of the options on the far left should be highlighted. If not, use the **[▼]** to place the blinking cursor on the appropriate entry and press **[ENTER]**. Exit **MODE** by pressing **[EXIT]**, **[CLEAR]**, or **[2nd] <QUIT>**. This accesses the home screen where expressions are entered.

☞ RETURN TO THE SECTION "INTEGER OPERATIONS" IN THE CORE UNIT (PG.2) AND CONTINUE UNTIL THE NEXT TI-86 PROMPT.

Absolute Value

The TI-86 does not have absolute value on its keypad. However, a key can be created using the **[CUSTOM]** key. Up to fifteen frequently used functions can be customized (and accessed with only two key strokes) provided the function desired is listed under **CATALOG**. To customize a function, press **[2nd] <CATLG/VARS>** followed by **[F3](CUSTM)**. Place the arrow next to **abs** in the list, then press **[F1].** The function **abs** is now listed under **PAGE↓**. The same procedure can be used to add other functions to the **CUSTOM** menu as necessary. Each time select an open slot in the menu by choosing the menu key below the open slot. Pressing **[MORE]** accesses additional slots for customizing. When finished, press **[EXIT]** until the menus at the bottom of the screen clear, and the home screen is displayed.

Pressing **[CUSTOM]** reveals the customized menu; to access a customized function, simply press the **F** key below the desired function. Pressing **[EXIT]** removes this menu.

Example: Simplify | -3 - 2 |
(Ensure you are at the home screen - press **[EXIT]** if necessary.)

Expression	**Verbally**	**Graphically**
| -3 - 2 |	The absolute value of the quantity negative three minus two.	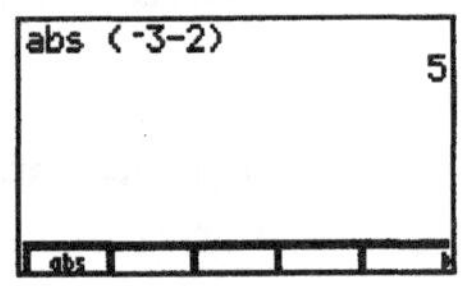 **Keystrokes:** **[CUSTOM] [F1](abs) [( ] [(-)] [3] [-] [2] [ )] [ENTER]** **Note:** Absolute value can also be accessed by pressing **[2nd] <MATH> [F1](NUM) [F5](abs).**

☞ RETURN TO THE "SQUARE ROOTS" SECTION OF THE CORE UNIT (PG.3) AND CONTINUE UNTIL THE NEXT TI-86 PROMPT. KEYSTROKES FOR YOUR CALCULATOR WILL DIFFER FROM THOSE GIVEN FOR THE TI-83/84 SERIES CALCULATORS; MAKE APPROPRIATE CHANGES IN THE MARGIN.

UNIT 2
FRACTIONS AND THE ORDER OF OPERATIONS

Prerequisite: Unit #1

TI-86	TI-86 USERS TURN TO "FRACTIONS AND THE TI-86" IN THE GUILDELINES (PG.17).

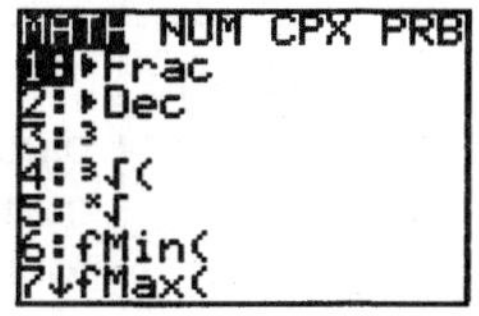

The calculator can be used to perform arithmetic operations with fractions. Pressing the **[MATH]** key reveals a math menu. Take a minute to use the **[▾]** cursor to scroll down the menu. Notice that there are 10 options available. Use of the first option, **1:▸ Frac**, will now be illustrated.

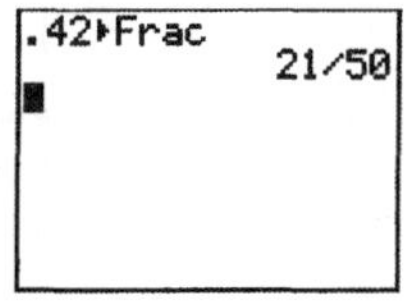

To convert 0.42 to a fraction enter the decimal number 0.42 by pressing **[.] [4] [2]**. Press **[MATH]** and under the displayed MATH menu press **[1](1:▸Frac)**. Press **[ENTER]** to activate the "convert to a fraction" command. Notice that the calculator displays the fraction reduced to lowest terms.

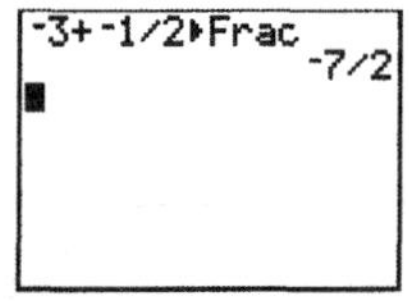

Recall that a mixed number is actually the sum of an integer and a fraction. Therefore, to enter mixed numbers on the TI-83/84 series indicate the addition of the integer and the fraction. To enter $-3\frac{1}{2}$, press **[(-)] [3] [+] [(-)] [1] [÷] [2] [MATH] [1] (1:▸Frac) [ENTER]**. Your calculator display should correspond to the one at the right. Because the entire fraction is negative, both the integer part and the rational part must be entered as negative numbers.

Order of Operations

The graphing calculator is programmed to follow the order of operations. The use of parentheses overrides the defined order. Use parentheses as a grouping symbol when using the graphing calculator.

Expression	Verbally	Analytically	Graphing Calculator	Notes
3 + 2 • 4	Three plus two times four.	3 + 2 • 4 = 3 + 8= 11	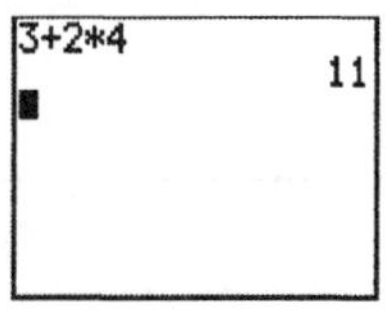	The multiplication must be done before the addition. The calculator follows the order of operations.
(3 + 2) • 4 OR (3 + 2)4	The quantity three plus two, multiplied by four.	(3 + 2) • 4 = 5 • 4 = 20	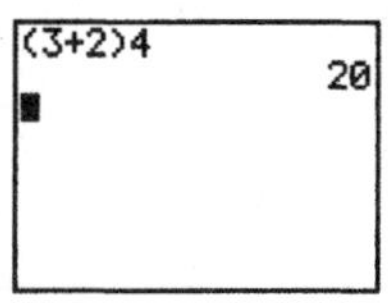	The parentheses override the order of operations and dictate that the addition be done before the multiplication. The use of a multiplication sign is optional as implied multiplication is programmed as such.

Expression	Verbally	Analytically	Graphing Calculator	Notes
3 + 5 - 2 + 4	Three plus five minus two plus four.	3 + 5 - 2 + 4 = 8 - 2 + 4 = 6 + 4 = 10	3+5-2+4 10	The calculator follows the order of operations and does not need parentheses to "tell" it how to evaluate the expression.
(3+5) - (2+4)	The quantity, three plus five, minus the quantity, two plus four.	(3+5) - (2+4) = 8 - 6 = 2	(3+5)-(2+4) 2	Parentheses are used to override the order of operations.
$\frac{6+3}{3}$	The quantity, six plus three, divided by 3.	$\frac{6+3}{3}=$ $\frac{9}{3}=$ 3	(6+3)/3 3	Parentheses must be used around the entire numerator because the fraction bar is written as a slash rather than a horizontal bar.
$6+\frac{3}{3}$	Six plus three divided by three.	$6+\frac{3}{3}=$ 6+1= 1 7	6+3/3 7	No parentheses are necessary because the numerator and the denominator both contain a single character.
3-{6 - [2 + 3]}	Three minus the quantity of 6 minus the quantity of two plus three.	3-{6 - [2 + 3]}= 3 - {6 - 5} = 3 - 1= 2	3-(6-(2+3)) 2	Only parentheses are used as grouping symbols on the calculator. It is programmed to follow the order of operations.

Example 1: Simplify $\frac{2}{3}+\frac{1}{5}$.

Expression	Verbally	Graphing Calculator	
$\frac{2}{3}+\frac{1}{5}$	The sum of two-thirds and one-fifth.	2/3+1/5▸Frac 13/15	**Keystrokes:** **[2] [÷] [3] [+] [1] [÷] [5] [MATH] [ENTER] [ENTER]**

NOTE: If a denominator is more than four digits, the decimal equivalent of the fraction will be returned. Enter the fraction $\frac{17}{10{,}000}$ and activate the convert to Frac option. The calculator will only display the decimal form of the number. ◆

Example 2: Simplify $\frac{2}{3} \div \frac{1}{5}$.

Expression	**Verbally**	**Graphing Calculator**
$\frac{2}{3} \div \frac{1}{5}$	Two-thirds divided by one-fifth.	 **Keystrokes:** **[( ] [2] [÷] [3] [ )] [÷] [( ] [1] [÷] [5] [ )] [MATH] [1] (1:▸Frac)[ENTER]**

Remember, the graphing calculator is programmed to follow the order of operations. Addition and subtraction problems involving fractions do not require the use of parentheses because the calculator will follow the order of operations and perform the division (fractions) operation first and then conclude with addition/subtraction. However, with division or multiplication, the operations are performed strictly from left to right following the order of operations. Therefore, to multiply or divide fractions, parentheses are necessary to override the order of operations.

Division ***by*** zero (0) is undefined:

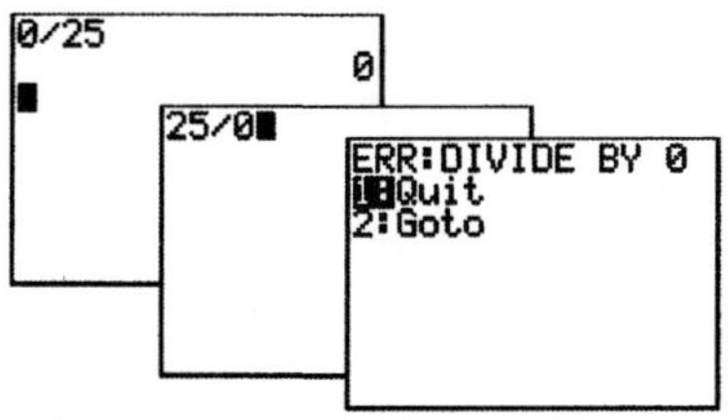

EXERCISE SET

Simplify the following expressions on the calculator; all answers should be expressed as fractions. **RECORD THE CALCULATOR SCREEN <u>LINE BY LINE EXACTLY</u>** as it appears.

1. $36 - 24 \div 3 + 5$

2. $24 \div 8 \bullet 2 - 4$

3. $28 - 3[5 + 2(7 - 2)]$

4. $\frac{5}{16} + \frac{1}{2} \bullet \frac{3}{8}$

5. $\frac{3}{5} \div \frac{1}{2} - \frac{5}{8}$

6. $\left|-\frac{3}{5}\right| \bullet \sqrt{\frac{1}{4}}$

7. $\left(\frac{1}{2}\right)^3 \bullet \frac{2}{3}$

8. $\sqrt{\frac{4}{25}} \div 2$

9. $\left(\frac{3}{5}\right)^2 \bullet \left(\frac{2}{3}\right)^3$

10. $\frac{5 + 3 \bullet 4}{(5 + 3)4}$

11. $\frac{5 - |1 - 3| + 7}{6 - 9 \div 3}$

12. $\frac{3 - |-15| \div 5}{-3(5 - |-2|) + 15}$

13. $\frac{6 + 12 \div 2 \bullet 3 - 5}{3^2 - 5^2 \bullet 3 + 6 \bullet 11}$

Note: Be sure to translate any error messages to the appropriate mathematical notation.

✍ 14. Enter the expression $\frac{16 + 8}{4}$ on your calculator in each of the following ways: 16 + 8 / 4 and (16 + 8)/4. Which is the correct format and why?

Solutions:

1. 33 **2.** 2 **3.** - 17 **4.** 1/2 **5.** 23/40 **6.** 3/10

7. 1/12 **8.** 1/5 **9.** 8/75 **10.** 17/32 **11.** 10/3 **12.** 0

13. Undefined **14.** The numerator must be grouped in parentheses, (16 + 8), in order for the calculator to divide the denominator, 4, into the entire numerator.

TI-86 GUIDELINES UNIT 2

Fractions and the TI-86

Since the ▸**Frac** option is used frequently, it will be added to the CUSTOM menu. To do this, press **[2nd] <CATALG/VARS> [F1] (CATLG) [F3](CUSTM)**. Cursor down the list to place the arrow next to ▸**Frac**, found near the bottom of the list. Press **[F2]** or another open **F** key to create the custom key. HINT: Since ▸**Frac** was found at the bottom of the list, pressing **[▲] [2nd] <M2>(PAGE↑)** until the arrow is positioned next to the desired selection would be quicker than scrolling through the entire list.

Note: To delete a customized entry, press **[2nd] [CUSTM] [F4](BLANK)** and then the **F** key that contains the command to be deleted.

Press **[EXIT]** until the blinking cursor is displayed at the HOME screen.

Enter the decimal number 0.42 by pressing **[.] [4] [2] [CUSTM] [F2](▸Frac) [ENTER]**. Notice that the calculator displays the fraction reduced to lowest terms.

Recall that a mixed number is actually the sum of an integer and a fraction. Therefore, to enter a mixed number, simply enter the expression as an indicated sum. To enter $-3\frac{1}{2}$ and express it as a fraction, press **[(-)] [3] [+] [(-)] [1] [÷] [2] [CUSTM] [F2](▸Frac).** Because the entire fraction is negative, both the integer part and the rational part must be entered as negative numbers. Your display should correspond to the one at the right.

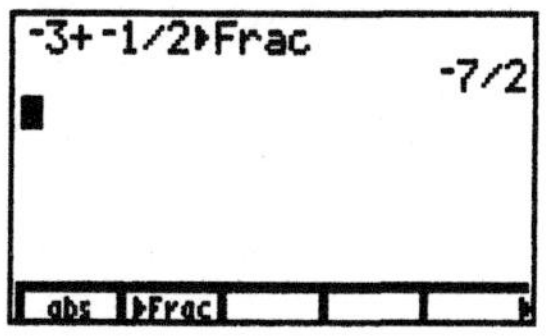

☞ Return to the section entitled "Order of Operations" in the core unit (pg.11) and complete the unit.

UNIT 3
EVALUATING THROUGH TABLES AND THE STOre FEATURE

*Prerequisite: Unit #2

To evaluate an expression means to find the value of the expression for given values of the variable(s). To analytically evaluate expressions, the substitution property is applied when constant values are substituted for specified variables. The expression is then simplified following the order of operations. The STOre feature and the TABLE of the calculator can sometimes be used to accomplish the same result.

Using the STOre Key

TI-86 IF USING THE TI-86, GO TO THE GUIDELINES (PG.27) WHICH FOLLOW THIS UNIT.

To store a value, enter the value, press **[STO▸]** and then the desired variable. For example, to store the value of 5 for x ($x = 5$), press **[5] [STO▸] [X,T,θ,*n*] [ENTER]**. The value 5 is now stored under the variable x. This value remains stored in x until another value replaces it. To check and see what value is stored under a specific variable, enter the variable at the prompt (blinking cursor) and then press **[ENTER].** The value is then displayed on the screen.

> NOTE: Because **X** is used as a variable so often in algebra, it has its own key on the calculator. The calculator is in function MODE; therefore when **[X,T,θ,*n*]** is pressed the screen displays the variable **X**.

Algebraically	Verbally	Graphically	Notation
Evaluate $3x^2 + 6x + 2$ when $x = -11$	Find the value of the polynomial when the variable x is replaced with the constant negative eleven.	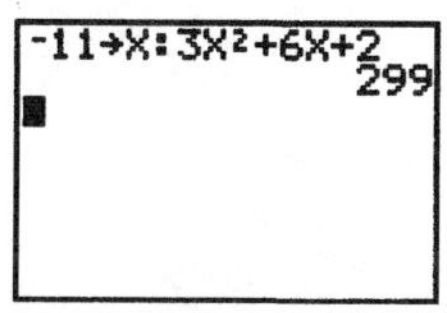 **Keystrokes:** **[(-)] [1] [1] [STO▸] [X,T,θ,*n*]** **[ALPHA] <:>** **[3] [X,T,θ, *n*] [x^2] [+] [6]** **[X,T,θ,*n*] [+] [2] [ENTER]**	• $x = -11$ is entered on the calculator as $-11 \rightarrow x$. •The colon ":" key (located above the decimal key) is used to separate commands that are to be entered on the same line. •Pressing ENTER activates the command.

To store a value under a variable other than **X**, the **[ALPHA]** key must be used to access the letters of the alphabet. Pressing the **[ALPHA]** key, followed by another key, allows access to the upper case letters and symbols written above the key pads in the same color as the alpha key.

Example 1: **Evaluate** $a^2 - 3a + 5$ when $a = -7$.

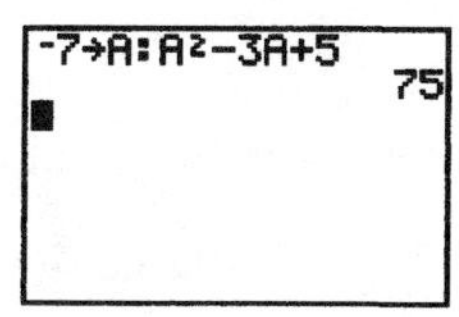

Solution: See screen at right.

*NOTE: Be sure to use the **[(-)]** key for the negative sign on a number and use the subtraction key, [-], to indicate subtraction.* ◆

EXERCISE SET

✍1. Look at the screen displayed at the right. What have you told the calculator to do?

```
-2→X:3→Y:X²-2X+Y
²
                 17
```

a. -2→X means ______________________________

b. 3→Y means ______________________________

c. $X^2 - 2X + Y^2$ means ______________________________

d. 17 means ______________________________

e. The purpose of the colon (:) is to______________________________

Exercises 2-6 should be completed as follows:

(a) evaluate each expression analytically by substituting the values for the variables into the given expression and following the order of operations, and

(b) evaluate with the calculator, using the **STO▸** feature. Record the screen display as a means of showing your work. You must copy the screen display **exactly** as it appears, line by line.

2. Evaluate $2x^2 + 3x + 1$ when $x = -3$.

Analytically:

Calculator display:

3. Evaluate $(x - 2)^2 + 3x$ when $x = \frac{1}{2}$.

Analytically:

Calculator display:

4. Evaluate $4(x - 3) + 5(y - 3) - 4$ when $x = -1$ and $y = -3$.

Analytically:

Calculator display:

5. Evaluate $-x^2 + 3xy^2 - 6y^3$ when $x = 3$ and $y = 4$.

TI-86 IF USING THE TI-86, GO TO THE GUIDELINES (PG.28) WHICH FOLLOW THIS UNIT.

Analytically: *Calculator display:*

6. Evaluate $\dfrac{2x^3y - 2x^2y + x}{5x - y}$ when $x = 2$ and $y = ½$.

Analytically: *Calculator display:*

7. Troubleshooting: The problem below has been entered **incorrectly** on the calculator. Make the necessary corrections so that the calculator display accurately represents the problem given. Verify with your calculator.

```
X→-3:2X-5█
```

Evaluate $2x - 5$ when $x = -3$.

✍ 8. Discuss the use of parentheses when entering expressions containing radicals or fractions.

Checking Solutions

The STOre feature of the graphing calculator can be used to check solutions to equations and to confirm that expressions are equivalent.

Example 2: Is 6 a solution to the equation $4(2x - 1) - 8 = 42 - x$?

```
6→X:4(2X-1)-8
                36
42-X
                36
```

Solution: If 6 is a solution to the equation then the left side, $4(2x - 1) - 8$, will have the same value as the right side, $42 - x$, when 6 is substituted for x. Store 6 in x and then find the value of each side of the equation. Your screen should look like the one at the right. ◆

A proposed solution can also be verified through the TABLE feature of the TI-83/84 series, or TI-86 graphing calculators.

Example 3: Consider the equation $3(5x + 1) = 8x + 24$. Is 3 a solution to this equation?
Solution: To confirm that 3 is a solution via the TABLE, we will *store* the expressions on each side of the equation on the **y-edit** screen. To do this, press **[Y=]** and CLEAR all entries.

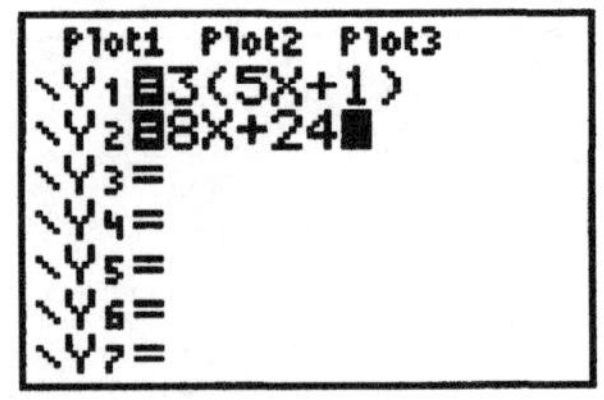

TI-86 PRESS **[GRAPH] [F1](Y(X) =)**.

Enter the left side of the equation at the **Y1=** prompt. Press **[ENTER]** to move the cursor to the **Y2=** prompt, which is where the right side of the equation will be entered. To display the appropriate TABLE value, press **[2nd] [TblSet]**.

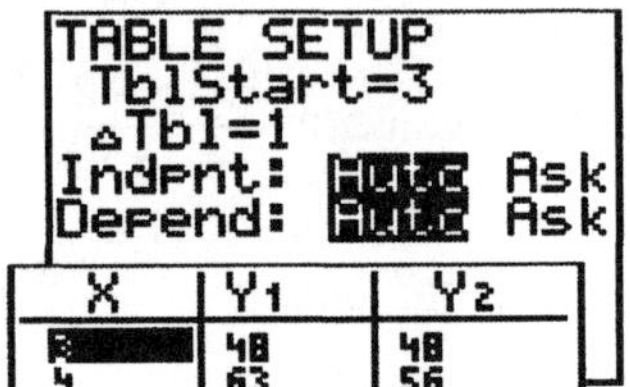

TI-86 TI-86 USERS PRESS **[TABLE] [F2](TBLST)**.

The cursor is now at the **TblStart=** prompt. Enter 3, since that is the value to be verified as a solution. Press **[ENTER]** to move the cursor to the **ΔTbl=** prompt. Enter a 1 (one) to increment the x-value of the TABLE by one unit. Since we will only be concerned with one entry line in the TABLE, the table increment is not critical. It is suggested that the increment be left at 1. Access the TABLE by pressing **[2nd] [TABLE]**.

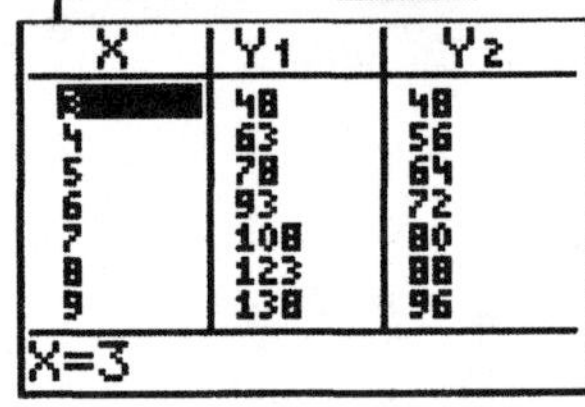

TI-86 TI-86 USERS PRESS **[F1](TABLE)**.

Your screen should look like the one pictured.

Notice the first entry under the X column is 3. Recall the expression $3(5x + 1)$ was entered at the Y1 prompt and $8x + 24$ was entered at the Y2 prompt. The values in the Y1 and Y2 column are equal (i.e. both 48) verifying that when $x = 3$, each of the expressions entered at the Y1 and Y2 prompts evaluate to 48.

Use the [▴] and [▾] cursors to scroll through the TABLE. Notice that the values in the Y1 and Y2 columns are equal ONLY when $x = 3$.

◆

The TABLE has provided numerical evidence that 3 is a solution to the equation $3(5x + 1) = 8x + 24$. In the space below, use the Substitution Property to evaluate both sides of the equation analytically (by hand) to confirm that $x = 3$ is a solution.

The **STO▸** feature can also be used to confirm that when 3 is substituted into the expression on each side of the equation, the expressions both evaluate to 48.

The STOre feature operates off the HOME SCREEN of the calculator. Therefore, any letter of the alphabet can be used as a variable. Use of the STOre feature requires you to store the *value of the variable* and then the expression to be evaluated.
Care must be taken when evaluating expressions using the TABLE feature. The *expression(s) is(are) stored* on the Y= screen. The *only* variable that can be used is *x*, and you *must* use the **[X,T,θ,*n*] ([x-var] on the TI-86)** to enter that variable.

STEPS FOR CONFIRMING A SOLUTION VIA THE TABLE

1. Press **[Y=]** and enter the left side of the equation at **Y1** and the right side at **Y2**.

TI-86	TI-86 USERS PRESS **[TABLE] [F2](TBLST).**

2. Press **[2nd] <TblSet>** to set the TblStart to the value believed to be the solution of the equation. The value for **ΔTbl** may be any value. It is suggested that the number one be used.

3. Access the TABLE by pressing **[2nd] <TABLE>** (TI-86 users press **[F1](TABLE).**) The solution is confirmed if and only if the Y1 and Y2 columns are identical for the designated value of X.

EXERCISE SET

Directions: Verify that each number below is a solution to the given equation. Copy the line(s) of the TABLE that verify the solution(s).

9. Verify that $\frac{3}{2}$ is a solution of the equation $x + 5 = 3x + 2$.

X	Y1	Y2

10. Verify that 2 is a solution of the equation $\frac{3}{2}(x+4) = \frac{20-x}{2}$.

X	Y1	Y2

11. Verify that 4 **and** -3 are both solutions of the equation $x^2 - x - 12 = 0$.

X	Y1	Y2

12. What two values of x can be verified as solutions to an equation from the displayed table?

X	Y1	Y2
-2	4	-1
-1	1	1
0	0	3
1	1	5
2	4	7
3	9	9
4	16	11

X= -2

x = _______ and x = ________

✍ What assures you that these numbers are solutions?

✍13. The solutions to the equation $3x^2 = 5x + 12$ are $-\frac{4}{3}$ and 3. When a student attempts to confirm the solution of $-\frac{4}{3}$, he instructs the calculator to display the TblStart as $-\frac{4}{3}$ and ΔTbl to be increased in increments of 1. Note that the TblStart of $-\frac{4}{3}$ is converted to a decimal approximation for display in the TABLE. The top row of the TABLE confirms $-\frac{4}{3}$ as a solution. How can the solution X = 3 be confirmed?

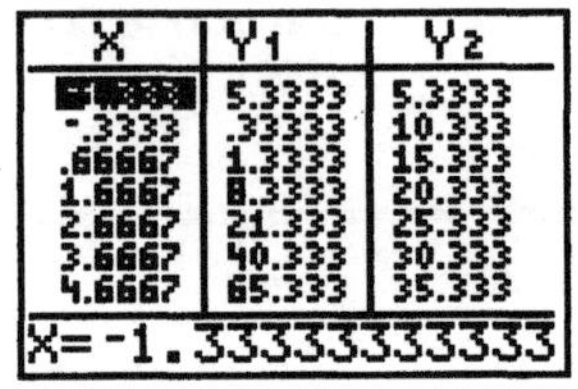

Example 4: Simplify the following expression algebraically: $3x(x - 2) + 5x^2$

Solution: The simplified expression is $8x^2 - 6x$. To determine if $3x(x - 2) + 5x^2$ equals $x^2 - 6x$, enter the original expression at the **Y1=** prompt and the simplified expression at the **Y2=** prompt. Set the TABLE to have a start value of 0 and increment the TABLE by 1. This start value and the increment are arbitrary. Any values could be used in either position. Your screen should match the one displayed.

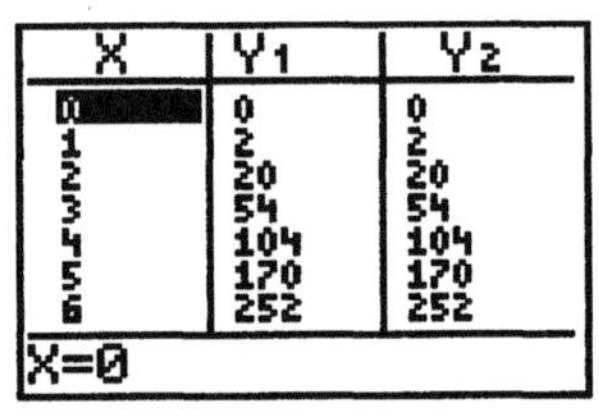

WARNING: This procedure confirms that two expressions are equivalent. It does not determine that the expression is completely simplified.

◆

EXERCISE SET CONTINUED

✍14. Explain **how**, in Example 3, the use of the TABLE confirms that the given expressions are equal.

✍15. The expression $(3x^2 - 5x + 2) - (8x^2 - 3x - 1)$ simplifies to $-5x^2 - 2x + 3$. Confirm that these two expressions are equivalent by using the TABLE feature of the graphing calculator. Describe the entries on the TABLE screen.

✍16. The expression $(4x - 5)(3x + 7)$ expands to $12x^2 + 13x - 35$. Confirm that these two expressions are equivalent by using the TABLE feature of the graphing calculator. Describe the entries on the TABLE screen.

17. In attempting to perform the division on the expression $\frac{4x^2-8x-16}{8x^2}$, a student produced

$\frac{1}{2}-\frac{1}{x}-\frac{4}{2x^2}$.

a. When he tried to use the calculator to confirm his work, he selected 0 as his TblStart for x and the calculator displayed an error message in **both** the Y1 and Y2 columns. What does this mean in terms of the two expressions?

b. Can he conclude that his simplification is correct?

18. A student expanded the quantity $(2x + 5)^2$ to $4x^2 + 25$. Confirm or deny with the TABLE feature that this is the correct expansion.

Solutions:

1. a. store -2 for the variable x **b.** store 3 for the variable y **c.** evaluate the expression for the stored value of the variables **d.** the value of the expression is 17 when $x = -2$ and $y = 3$

e. separate commands

2. 10 **3.** 15/4 **4.** -50 **5.** -249 **6.** 12/19

7. Your screen display should indicate that -3 is stored in X: -3→X

8. TI-83/84 Series: Because the calculator automatically inserts a left parenthesis when the square root symbol is accessed, students must remember to close the parenthesis appropriately.
TI-86: Parentheses should be used around expressions under the radical sign that contain more than one term.

Parentheses should be used around fractions when multiplying or dividing to ensure the correct application of the order of operations (for all series of calculators).

9.

X	Y1	Y2
1.5	6.5	6.5

10.

X	Y1	Y2
2	9	9

11.

X	Y1	Y2
-3	0	0
4	0	0

12. $x = -1$, $x = 3$; The Y columns are equal for these X values.

13 Change the table minimum (start) to 3.

14. Entry by entry, all values of Y1 equal all values of Y2.

15. Entry by entry, all values of Y1 equal all values of Y2.

16. Entry by entry, all values of Y1 equal all values of Y2.

I7a. The expression is undefined when X=0.

I7b. It is correct, but not completely simplified.

18. The expansion is not correct because the TABLE values for Y1 and Y2 are not equal.

TI-86 GUIDELINES UNIT 3

The **[x-VAR]** key is used to display the variable **x**. Note that the variable **x** is displayed as a lowercase letter on the TI-86, instead of an uppercase **X** as the **[X,T,θ,*n*]** key on the TI-83/84 series. When using this text, each **X** printed as a variable for the TI-83/84 series should appear as **x** on the TI-86.

To store a value, simply enter the value, press **[STO▸]** and then the desired variable. For example, to store $x = 5$, press **[5] [STO▸] [x-VAR] [ENTER]**. The value 5 is now stored under the variable **x**. This value will remain stored in **x** until another value replaces it. To determine the value that is actually stored under a specific variable, enter the variable at the prompt (blinking cursor) and then press **[ENTER].** The value is then displayed on the screen.

NOTE: Pressing the **STO▸** key automatically initiates the ALPHA key, allowing access to the remaining letters of the alphabet. Press **STO▸** and observe that the blinking cursor has an **A** in it for **ALPHA.** Letters will automatically be recorded in uppercase format. To disengage the ALPHA feature, press the **[ALPHA]** key.

Algebraically	**Verbally**	**Graphically**	**Notation**
Evaluate $3x^2 + 6x + 2$ when $x = -11$	Find the value of the polynomial when the variable *x* is replaced with the constant negative eleven.	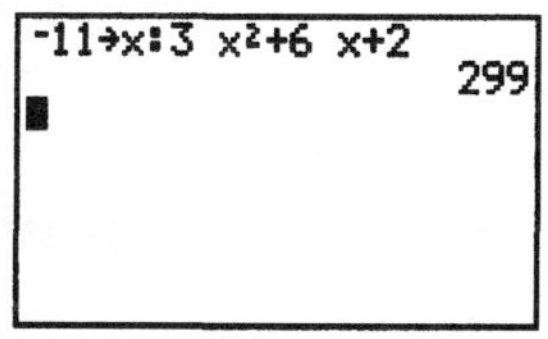 **Keystrokes:** **[(-)] [1] [1] [STO▸] [x-VAR]** **[2nd] <:> [ALPHA]** **[3] [x-VAR] [x²] [+] [6]** **[x-VAR] [+] [2] [ENTER]**	• $x = -11$ is entered on the calculator as $-11 \rightarrow x$. •The colon **":"** key (located above the decimal key) is used to separate commands that are to be entered on the same line. • The blinking cursor has an **A** in it for ALPHA. ALPHA was engaged by the STOre key; disengage the ALPHA feature by pressing **[ALPHA]** . •Pressing ENTER activates the command.

To store a value under a variable other than *x*, the **[ALPHA]** key must be used to access the letters of the alphabet. Pressing the **[ALPHA]** key, followed by another key, allows access to the upper case letters and symbols written above the key pads in the same color as the alpha key.

◆

The TI-86 has both uppercase and lowercase alphabet letters that can be used as variables. To access a lowercase letter when using the **STO▸** feature, press **[2nd] <ALPHA>** after pressing the **[STO▸]** key. However, realize that an expression containing a lower case variable could be entered as an expression using an upper case variable as long as you are consistent. Use lower case letters with care. As noted at the end of this guideline, some lower case letters are reserved for system variables.

When not using the **STO▸** feature, uppercase letters are accessed by **[ALPHA]** and lowercase letters are accessed by **[2nd] <ALPHA>**.

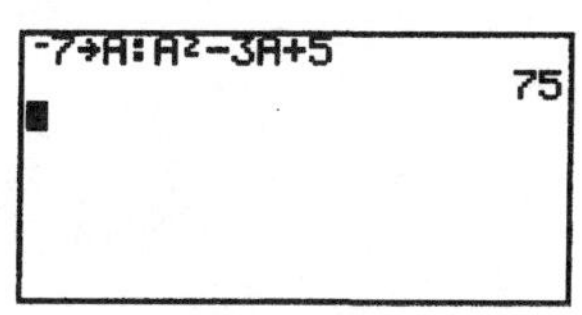

Example 1: **Evaluate** $a^2 - 3a + 5$ when $a = -7$.

Solution: See screen at right.

*NOTE: Be sure to use the gray **[(-)]** key for the negative sign on a number and use the black **[-]** key to indicate subtraction.*

The keystrokes would need to be **[(-)] [7] [STO▸] [A] [2nd] <:> [A]**(*observe that the cursor is still blinking with an uppercase A inside of it*) **[ALPHA]** (*the ALPHA feature is now disengaged*) **[x^2] [-] [3] [ALPHA] [A] [+] [5] [ENTER].**

◆

☞ RETURN TO THE EXERCISE SET IN THE CORE UNIT (PG.20) AND WORK THE EXERCISES.

The TI-86 has the capability of recognizing combinations of letters as independent variables. For example, the expression **4AC** means "four times A times C". However, the TI-86 would recognize **AC** as representing only **<u>one</u>** unknown, not a product of two unknowns. Thus the expression "four times A times C" must be displayed on the TI-86 as **4A*C** or **4A C**. Variables can be separated with a space **[␣]** to denote the multiplication of two different quantities. By being able to combine letters to form new variables, the TI-86 has a limitless list of variables available for use. Uppercase letters are reserved for user variable names. Note that the TI-86 would recognize **4a*c** as an entirely different product since the variables are lowercase. Thus Exercise #5 on page 21 should be entered with either a multiplication sign or a **space** between the two variables in the middle term; i.e. $3x * y^2$ or $3x\ y^2$, but not $3xy^2$.

Lowercase letters should be used with care as some are reserved for system variables. Examples of these are:

a	coefficient of regression
b	coefficient of regression
c	speed of light
e	e natural log base
g	force of gravity
h	Planck's constant
k	Boltzman's constant
n	number of items in a sample
u	atomic mass unit

The variables r,t,x,y, and θ are updated by graph coordinates based on the graphing mode.

☞ RETURN TO THE CORE UNIT EXERCISE 5 (PG.21) AND COMPLETE THE UNIT.

UNIT 4
RATIONAL EXPONENTS AND RADICALS

*Prerequisite: Unit #2

TI-83/84users should press **[MODE]** and verify that **Real** is highlighted in the left column. The TI-83/84series calculators have complex number capabilities as is indicated by the "a + bi" selection adjacent to **Real**. Complex number operations will be discussed in the unit entitled *TI-83/84 Series and TI-86: Complex Numbers*. For now, the calculator should be set to compute only with real numbers.

Radicals

The inverse operation of raising a number to a power is extracting a root. For example, $4^3 = 64$ and $\sqrt[3]{64} = 4$.

The only type of radical that has been addressed thus far in this text is $\sqrt{\;}$ or principal square root. We will now examine $\sqrt[3]{\;}$, $\sqrt[4]{\;}$, $\sqrt[6]{\;}$, , etc., and the relationship of these radicals to rational exponents.

TI-86 PRESS **[2ND] <MATH>** TO DISPLAY THE FIRST FIVE SUB-MENUS OF THE TI-86'S MATH FEATURE. THE "▸" AFTER **MISC** INDICATES THERE ARE MORE SUB-MENUS. PRESS THE **[MORE]** KEY TO DISPLAY THE REMAINING MENU AND THEN PRESS **[MORE]** ONE MORE TIME TO RETURN TO THE INITIAL SET OF SUB-MENUS. (SUGGESTION: AFTER COMPLETING THIS UNIT, EXAMINE THE CONTENTS OF EACH OF THESE SIX SUB-MENUS WHICH ARE ACCESSED BY PRESSING THE APPROPRIATE **F** KEY.) THE RADICAL SYMBOL, $\sqrt[x]{\;}$, THAT WILL BE USED IN THIS UNIT IS FOUND UNDER THE MISC SUBMENU. PRESS **[F5](MISC)** FOLLOWED BY **[MORE]** TO DISPLAY THE $\sqrt[x]{\;}$. AT THIS POINT, IT WOULD BE A GOOD IDEA TO RETURN TO THE CUSTOM MENU AND CUSTOMIZE THE $\sqrt[x]{\;}$ USING THE CATALOG.

THERE IS NO $\sqrt[3]{\;}$ AVAILABLE ON THE TI-86; CONTINUE THE UNIT ON THE NEXT PAGE.

Begin by looking at the MATH menu. Press **[MATH]** and the appropriate screen is displayed at the right. To scroll through the entire menu, press the down arrow key. This unit will use the **▸FRAC** option, as well as the $\sqrt[3]{\;}$ and $\sqrt[x]{\;}$ options.

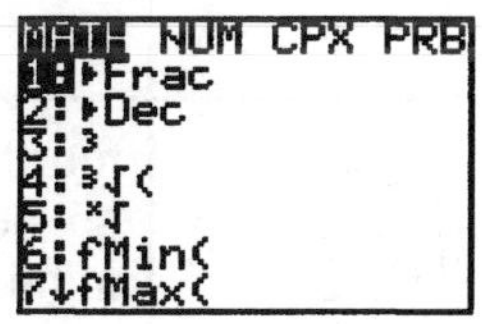

Expression	Verbally	Graphically	Notation
$\sqrt[3]{-27}$	the cube root of negative twenty-seven	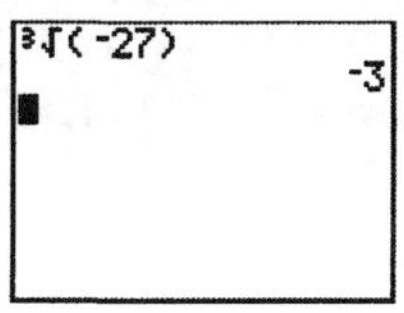 **Keystrokes:** **[MATH] [4] (4: $\sqrt[3]{(}$) [(-)] [2] [7] [ )] [ENTER]**	The calculator automatically enters a left parenthesis when the cube root feature is accessed. Make sure you close the parenthesis around the radicand.

Expression	**Verbally**	**Graphically**	**Notation**
$7\sqrt[3]{5}$	seven times the cube root of five	7³√(5) 11.96983163 **Keystrokes:** **[7] [MATH] [4] (4: $\sqrt[3]{(}$)** **[5] [ENTER]**	• The result displayed is an approximate answer rounded to eight decimal places: 11.96983163. • Notice that the calculator recognizes implied multiplication.
$\sqrt[6]{64}$	the sixth root of sixty-four	6 ˣ√64 2 **Keystrokes:** **[6] [MATH] [5] (5: $\sqrt[x]{\;}$) [6] [4] [ENTER]**	Designate a value for x by entering the index first and then the $\sqrt[x]{\;}$ symbol.
$4\sqrt[6]{64}$	four *times* the sixth root of sixty-four	4*6 ˣ√64 8 **Keystrokes:** **[4] [X] [6] [MATH]** **[5] (5: $\sqrt[x]{\;}$) [6] [4]** **[ENTER]**	Implied multiplication cannot be used because of the calculator notation. Had the expression $4\;6\sqrt[x]{64}$ been entered, the calculator would compute $\sqrt[46]{64}$. A multiplication symbol, " * ", must be entered after the digit 4 for clarification.
$\sqrt[4]{\dfrac{16}{625}}$	fourth root of sixteen six hundred twenty-fifths	4 ˣ√(16/625)▸Frac 2/5 **Keystrokes:** **[4] [MATH] [5] (5: $\sqrt[x]{\;}$)** **[( ] [1] [6] [÷] [6] [2] [5]** **[ )] [MATH] [1] (1:▸Frac)** **[ENTER]**	The correct root is 2/5 (or 0.4). If you got 2/625 (or 0.0032) then you failed to group the fraction with parentheses. Parentheses are necessary for the calculator to extract the fourth root of the quantity 16/625.

EXERCISE SET

Directions: Evaluate each radical expression. Record the screen display, being sure that the indicated root is displayed as an integer or fraction - rather than a decimal.

1. $\sqrt[3]{-125}$

2. $\sqrt[4]{4096}$

3. $\sqrt[6]{5^6}$

4. $\sqrt{\dfrac{4}{25}}$

5. $\sqrt[5]{\dfrac{1}{32}}$

6. $\sqrt{-625}$

Rational Exponents

Rational exponents are entered into the calculator in the same manner as positive and negative integer exponents. Be aware of the order of operations as you enter the expression. An expression has been entered on the calculator and is displayed on the first screen at the right. Two operations are indicated, raising a base to a power and division. Because of the order of operation rules, the power will be performed before the division. The first calculator display is not an illustration of $25^{\frac{1}{2}}$ but of $\dfrac{25^1}{2}$. For the calculator to evaluate the expression $25^{\frac{1}{2}}$, parentheses must be inserted to override the order of operations as displayed on the second screen.

```
25^1/2
```

```
25^(1/2)
```

EXERCISE SET

The following problems have been placed in groups of four to more easily compare the effect of differing exponents on the same base. **Be sure all rational exponents are enclosed in parentheses.** Enter final results on the appropriate blank.

7. $25^2 =$ ____ $25^{1/2} =$ ____ $25^{-1/2} =$ ____ $25^{-2} =$ ____

8. $8^3 =$ ____ $8^{1/3} =$ ____ $8^{-1/3} =$ ____ $8^{-3} =$ ____

9. $\left(\dfrac{16}{81}\right)^2 =$ ____ $\left(\dfrac{16}{81}\right)^{\frac{1}{2}} =$ ____ $\left(\dfrac{16}{81}\right)^{-\frac{1}{2}} =$ ____ $\left(\dfrac{16}{81}\right)^{-2} =$ ____

The following exercises allow you to determine the relationship between $\sqrt[n]{a^m}$ and $a^{m/n}$. The problems are grouped in such a way that patterns can be discovered. Record the result on the blank.

10. $\sqrt[3]{4^6}$ = _____ $\quad 4^{\frac{6}{3}}$ = _____ $\quad 4^2$ = _____

11. $\sqrt[3]{-8}$ = _____ $\quad (-8)^{\frac{1}{3}}$ = _____ $\quad \sqrt[3]{(-2)^3}$ = _____

12. $\sqrt{16}$ = _____ $\quad 16^{\frac{1}{2}}$ = _____ $\quad \left(\frac{1}{16}\right)^{-\frac{1}{2}}$ = _____

13. $27^{\frac{2}{3}}$ = _____ $\quad \left(\sqrt[3]{27^2}\right)$ = _____ $\quad \sqrt[3]{27^2}$ = _____

14. $\sqrt{-36}$ = _______ $\quad (-36)^{\frac{1}{2}}$ = _______

✍15. In #14 the TI-83/84series calculators displayed a DOMAIN ERROR. Explain what is wrong with the problem.

✍16. Explain the relationship between $\sqrt[n]{a^m}$ and $a^{m/n}$.

Solutions: **1.** -5 **2.** 8 **3.** 5 **4.** 2/5 **5.** ½

6. Did you get an error message? The DOMAIN error message is displayed because you were instructed to set the calculator to compute only with real numbers. In the real number system, $\sqrt{-625}$ does not exist. There is no real number that when raised to the second power will equal -625.
The TI-86 will display (0,25) instead of an ERROR message as the TI-83/84 series do. This is because the TI-86 can perform complex number operations. The $\sqrt{-625}$ simplifies to the complex number 0 + 25i. The calculator notation for 0 + 25i is (0,25). The use of the TI-86 in the simplification of complex number expressions is discussed in detail in the unit entitled *TI-83/84 Series and TI-86: Complex Numbers.* Until the instructor and/or textbook addresses complex numbers, results displayed in the form (a, b) should be recorded as *no real number.*

7. 625, 5, 1/5, 1/625 **8.** 512, 2, 1/2, 1/512 **9.** 256/6561, 4/9, 9/4, 6561/256
10. 16, 16, 16 **11.** -2, -2, -2 **12.** 4, 4, 4 **13.** 9, 9, 9
14. not a real number, not a real number

TI-86	TI-86 USERS WILL HAVE (0,6) FOR BOTH RESULTS IN 14. REMEMBER, THIS IS THE CALCULATOR'S NOTATION FOR COMPLEX NUMBERS WHICH WILL BE ADDRESSED IN A LATER UNIT.

15. The square root function is not defined for negative radicands.

16. In the fractional exponent $\frac{m}{n}$, the *n* designates the index of the radical and the *m* is the exponent applied to the radical.

UNIT 5
TI-83/84 Series and TI-/86: COMPLEX NUMBERS

*Prerequisite: Unit #3

Complex numbers are numbers that can be expressed in the form a + bi where **a** represents the real part and **b** represents the imaginary part.

TI-83/84series The TI-83/84 series have the value "i" on the calculator face, located above the decimal point. This allows the easy entry and arithmetic manipulation of complex number expressions.

1. Complex numbers can be represented by variables, that is to say they can be stored as variables. Let $x = 4 - 2i$ and evaluate $3x^2 + 2x - 5$. Using the **STO▸** key, your display should look like the one at the right. The interpretation of the display is that replacing x with the complex number 4 - 2i in the expression $3x^2 + 2x - 5$, yields a value of 39 - 52i.

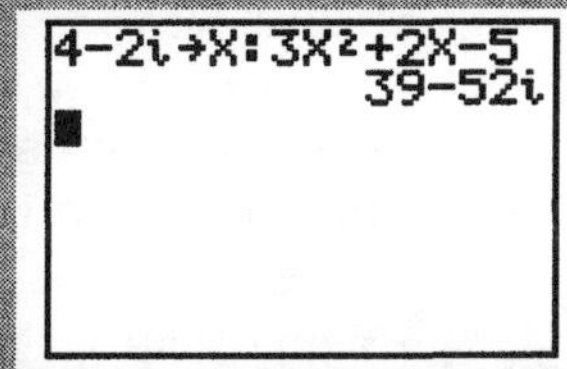

2. Access the complex number menu by pressing **[MATH]** [▸] [▸] to highlight **CPX**. This menu indicates that for any complex number, a + bi, the calculator will identify the conjugate, determine the real part, determine the imaginary part, and determine the absolute value. The **▸FRAC** option can be used with any of the operations.

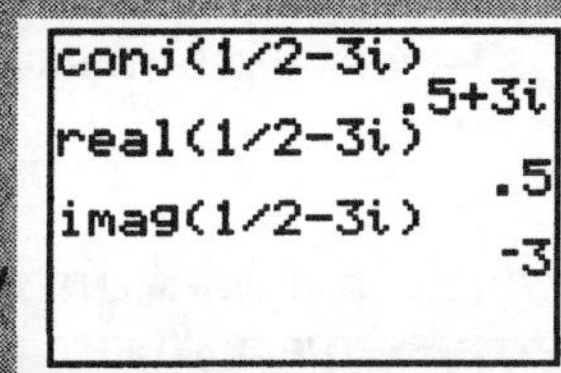

3. Arithmetic operations with complex numbers can be performed easily with these calculators. They are programmed to rationalize denominators. To perform the division $\frac{2 + 3i}{3 + 2i}$ be sure to use parentheses around both the numerator and denominator when entering the fraction and use the **▸FRAC** option to express the quotient in fraction form. The displayed result should correspond to the one pictured.

TI-86 COMPLEX NUMBERS ARE DISPLAYED AS **(REAL, IMAGINARY)** ON THE TI-86. THUS 2 - 3i WOULD BE DISPLAYED AS (2,-3) SINCE 2 IS THE REAL PART AND -3 IS THE IMAGINARY PART.

NOTE: (REAL, IMAGINARY) IS RECTANGULAR FORMAT, NOT POLAR

IN THE UNIT ENTITLED "EVALUATING THROUGH TABLE AND THE STORE FEATURE", PG.28, THE USE OF LOWERCASE LETTERS AS SYSTEM RESERVED VARIABLES IS DISCUSSED. THE LOWERCASE **i** CAN BE DESIGNATED AS A USER DEFINED VARIABLE IN THE TI-86 BY USING THE CONSTANT EDITOR. USE OF THE CONSTANT EDITOR TO DEFINE **i** AS (0,1), REPRESENTING 0 + i, PREVENTS YOU FROM STORING AN ALTERNATE VALUE IN **i** AT A LATER DATE.

Press **[2nd]** **<CONS>** **[F2](EDIT)** and enter a lowercase **i** after **Name=** by pressing **[2nd]** **<ALPHA>** <i>. Cursor down to **Value=** and enter **(0,1)** to define the value as the complex number 0+i. Complex numbers may now be entered on the home screen in a + bi format instead of (a,b) format if desired. However, answers will be returned by the calculator in the (a,b) format.

1. Complex numbers can be represented by variables, that is to say they can be stored as variables. Let x = 4 - 2i and evaluate $3x^2 + 2x - 5$. Using the **STO▸** key, the display should look like the one at the right. Interpreting the display, we see that when x = 4 - 2i, $3x^2 + 2x - 5$ has a value of 39 - 52i.

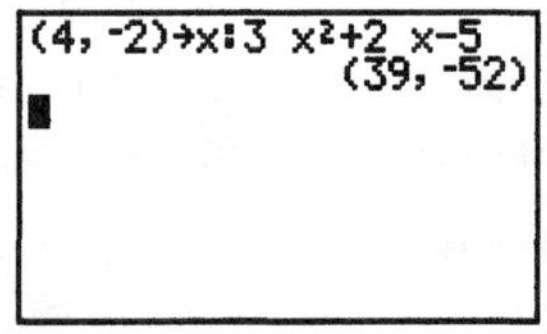

2. Press **[2nd]** **<CPLX>** to access the complex number menu. For any complex number, a + bi, the calculator will identify the conjugate (F1), determine the real part (F2), determine the imaginary part (F3), and determine the absolute value (F4). The remaining options are used with polar format.
The displays at the right indicate the applications of each of these options when applied to ½ - 3i. The **▸FRAC** option can be used with any of these operations.

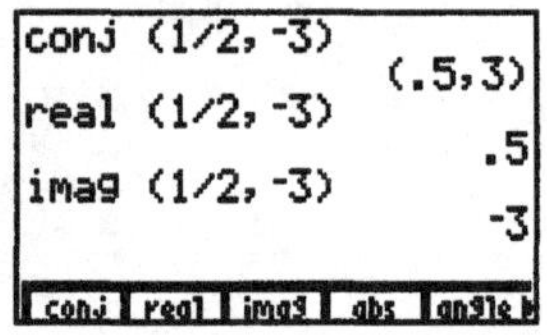

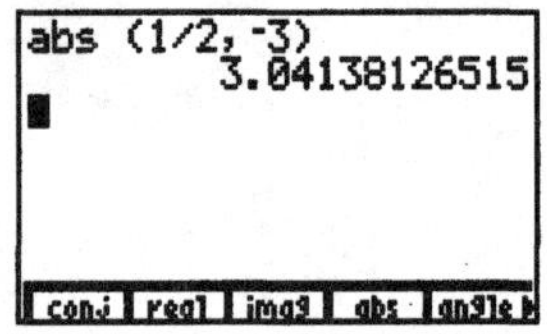

3. Arithmetic operations with complex numbers can be performed easily with the TI-86. It is programmed to rationalize denominators. To perform the division $\frac{2+3i}{3+2i}$ be sure to use parentheses around both the numerator and denominator when entering the fraction and use the **▸FRAC** option to express the quotient in fraction form. The displayed result is $\frac{12}{13}+\frac{5}{13}i$ in standard form.

EXERCISE SET

Directions: Use the complex number format and menu on the calculator to perform the complex number operations below. Copy your screen display and record all answers in standard a + bi form.

1. (3 + 2i) + 4(-2 + 5i)   1.________

2. (3 + 2i)(-2 + 5i)   2.________

3. $\frac{5+i}{2-3i}$ 3.________

4. $(2 - 3i)^3$ 4.________

5. i^{23} 5.________

6. i^{114} 6.________

7. Use the **STO▸** feature to show that $\pm 5i\sqrt{2}$ is a solution to $x^2 + 50 = 0$.

8. $\left|\frac{4}{5}+\frac{2}{3}i\right|$ 8.________

Can the result be converted to a fraction?____________

✍ Compute this same problem by hand and explain why.

> **TI-83/84 series** TI-83/84series users should enter $\sqrt{-4}$ on their calculator. If the **MODE** is still set to highlight **Real** then the calculator will display an error message: nonreal answer. At this point change the **MODE** setting to highlight **a + bi** so the calculator will compute the value of $\sqrt{-4}$. The result should now be 2i. When computing even roots of negative numbers, the calculator's MODE must be set for complex numbers.

9. Simplify the following on the calculator:

a. $4+\sqrt{-4}$ 9a.________

b. $\frac{2}{5}+\sqrt{-\frac{4}{9}}$ (Be sure the answer is expressed in fractional form.) 9b.________

10. Evaluate $\sqrt{-2}$ on the calculator. Express the result as a fraction, or explain why the calculator will not convert it to a fraction.

Solutions: **1.** $-5+22i$ **2.** $-16+11i$ **3.** $\frac{7}{13}+\frac{17}{13}i$ **4.** $-46 - 9i$

5. TI-83: 3E -13 -i which is equivalent to 0 - *i*
TI-86: (0,-1) which is 0 - *i* in standard form

6. TI-83: -1 + 1.4E -12 which is equivalent to -1 + 0i or -1
TI-86: (-1,0) which is -1 + 0*i* in standard form or -1.

7.

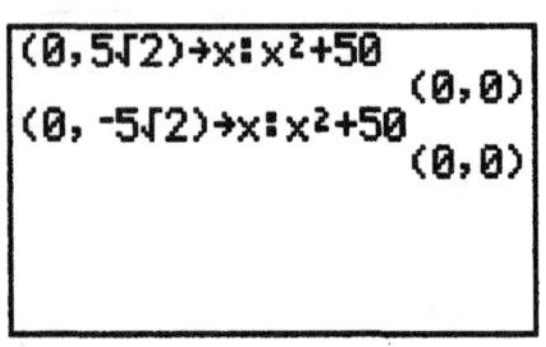

TI-86 Screen

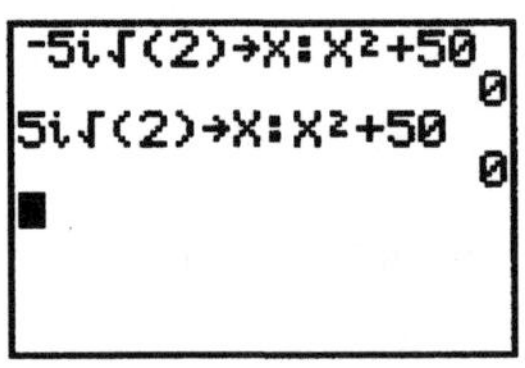

TI-83 Plus Screen

8. 1.04136662345: This number cannot be converted to a fraction because it is an irrational number.

9a. $4+2i$ **9b.** $\frac{2}{5}+\frac{2}{3}i$ **10.** The number cannot be converted to a fraction because it is an irrational number.

UNIT 6
GRAPHICAL SOLUTIONS: LINEAR EQUATIONS

*Prerequisite: Unit #3 is a prerequisite for this unit.
Note: Depending on the organization of your textbook you may choose to work units 14-15 prior to this unit.

Conditional Equations

This unit explores the use of the INTERSECT feature for solving equations whereas Unit 8 explores the use of the ROOT/ZERO feature. Recall, an equation is a statement that two algebraic expressions are equal. A solution is the value of the variable that makes the equation true.

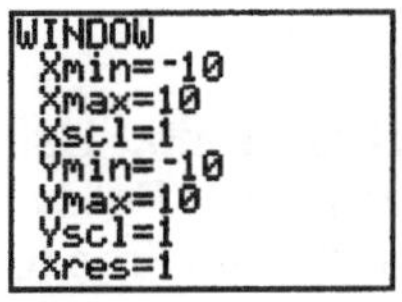

The INTERSECT option will be used to graphically find solutions to equations. It is not dependent on the window values entered. The only requirement is that the point of intersection be visible on the screen. The set of window values displayed to the right is designated as the standard viewing window . In order to maintain a consistent point of reference, all graphical solutions to equations will begin by displaying the graph in the standard viewing window. This can be easily set by pressing **[ZOOM] [6] (6:ZStandard).** (TI-86 users press **[GRAPH] [F3](ZOOM) [F4](ZSTD).**)

Check your WINDOW values to ensure they are the same as those displayed. Use your ZOOM menus (as specified above), or press **[WINDOW]** and use the down arrow key to move down the screen and change the values one by one.

TI-86	TI-86 USERS GO TO THE GUIDELINES WHICH FOLLOW THIS UNIT (PG.43).

Verbally	**Algebraically**	**Numerically**
Find the solution(s) to the equation 5x - 1 = - 3.	$5x-1=-3$ $5x=-2$ $x=-\frac{2}{5}$	$5\left(-\frac{2}{5}\right)-1=-3$ $-2-1=-3$ $-3=-3$ 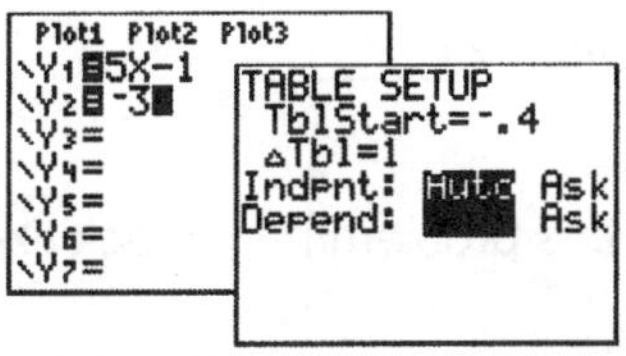 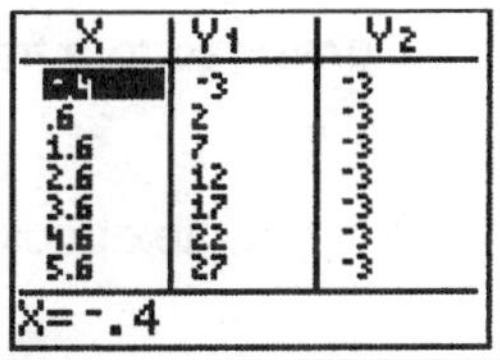 This is a verification of the algebraic solution.

Graphically	**Notation**
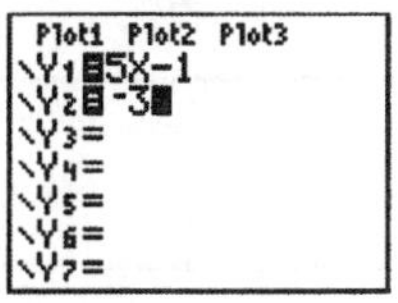 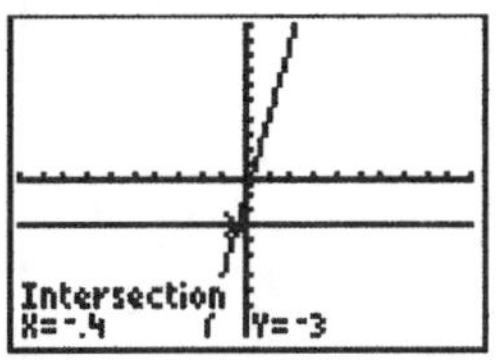 **Keystrokes:** • access INTERSECT option: **[2nd] <CALC>** (**CALC** is located above TRACE.) • **[5] (5:intersect)** • Move the cursor along the first curve to the approximate point of intersection; press **[ENTER]** • At the *second curve* prompt, press **[ENTER]**. • At the *guess* prompt, press **[ENTER]**. You do not have to guess because there is only one point of intersection. The solution is -0.4.	• The solution, as read from the graphical display, is $x = -.4$. • The number - 0.4 is equivalent to the fraction $-\frac{2}{5}$. To convert the decimal to a fraction, from the INTERSECTION screen, return to the home screen (**[2nd] <QUIT>**) and convert the value stored at X to a fraction (**[X,T,θ,*n*] [MATH] [1] (1:▸Frac) [ENTER]**). 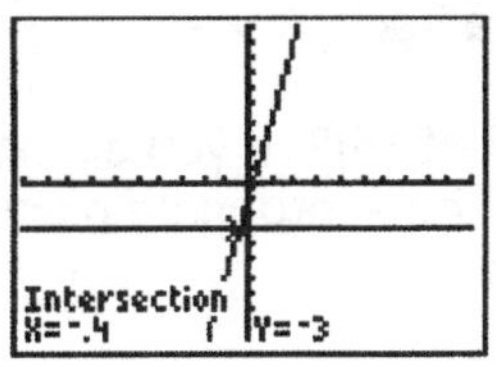 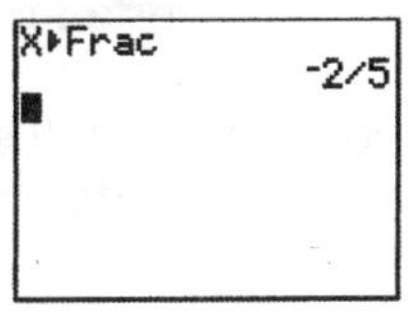

Steps for Solving An Equation Using the INTERSECT Feature

1. Enter the left side of the equation at the Y1 prompt on the *y*-edit screen and the right side at the Y2 prompt.
2. Make sure the calculator is in the standard viewing window.
3. Look at the graph and make sure that the point of intersection is clearly visible on the screen. Adjust the viewing window if necessary - see subsequent section.
4. Access the INTERSECT feature and respond to the prompts.
5. The solution to the equation is the *x*-value displayed on the screen.

EXERCISE SET

Directions: Using the INTERSECT option, solve these equations graphically (use the standard viewing window). Follow the same procedure as outlined. Sketch the INTERSECT screen that yields the solution and use ▸Frac to convert all decimal answers to fractions.

1. $5 = 2 - 7x$

   x = ______

   Converted to a fraction, x = ____

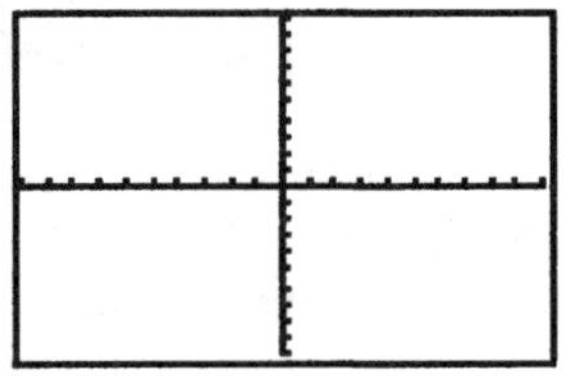

2. $(-4/3)x = -2$

   x = ______

   Converted to a fraction, x = ____

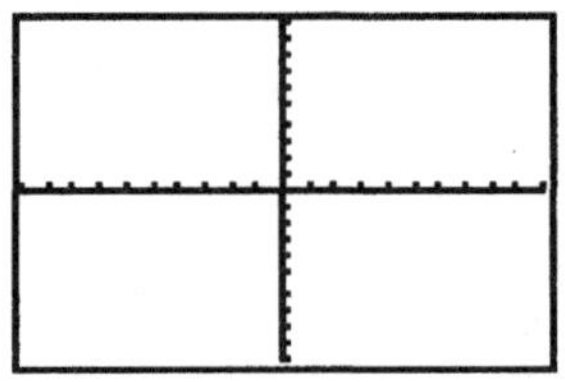

   *(In your textbook, this problem might look like this: $-\frac{4}{3}x = -2$)

3. $\dfrac{-4x}{3} + 6 = -1$

   x = ______

   Converted to a fraction, x = ____

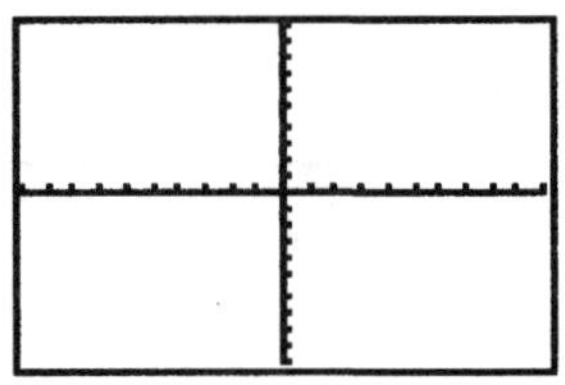

4. $\dfrac{2x-1.2}{0.6} = \dfrac{4x+3}{-1.2}$

   x = ______

   Converted to a fraction, x = ____

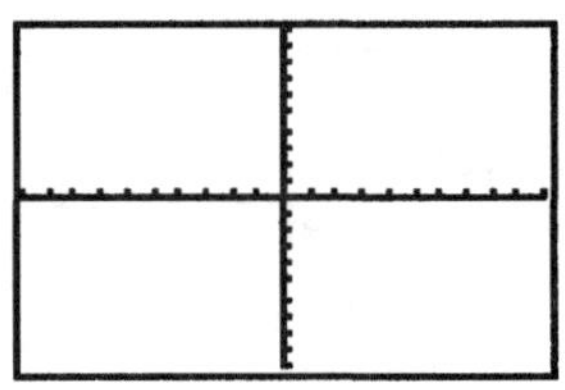

Did you get -3/40 (i.e. -.075)? If you did not, check the way the equation was entered. Parentheses will need to be inserted in the appropriate places.

5. When the equation 7x + 8 = -9 - 3x is solved graphically, the bottom of the screen displays the message "INTERSECTION x = -1.7 y = -3.9". Explain the meaning of the *x* and *y* values in the context of the equation that was being solved.

Adjusting Windows

Recall that the axes seen displayed on the graph screen are simply two number lines placed perpendicularly at their origins. The horizontal number line is the X-axis and the vertical number line is the Y-axis. They are oriented so that right is the positive and up is also the positive direction. The term Xmax on the WINDOW screen (WIND on the TI-86) refers to the maximum distance from the origin (0) to the right side of the viewing window, whereas Xmin refers to the distance from the origin to the left side of the window. Ymax is the distance from the origin to the top of the viewing window, whereas Ymin is the distance from the origin to the bottom of the viewing window.
It is best that the point(s) of intersection be visible in the viewing window if using the INTERSECT option. When not visible, one or both of the axes can be stretched.

Example 1: Solve the equation 3(*x* - 4) = 6 + *x* graphically, using the INTERSECT feature.

Solution: After entering the left and right sides of the equation at the appropriate prompts and pressing **[GRAPH]** , the displayed graph appears. The viewing WINDOW is not sufficiently large to allow the display of the point of intersection.

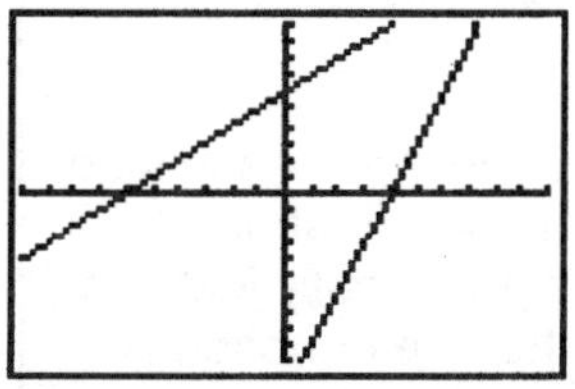

Which part of which axis must be stretched? by how much?

The Ymax must be made larger, and it is suggested that you simply make an educated guess. After accessing the WINDOW menu, change Ymax to 20. The point of intersection should be clearly visible as illustrated by the second screen. Use of the INTERSECT option displays the correct solution of 9. ◆

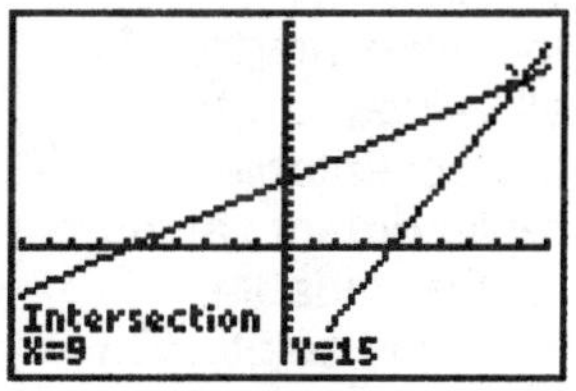

EXERCISE SET

Directions: Solve each equation using the INTERSECT feature of the calculator. Sketch the graph displayed, including the axes. Record window values in the spaces provided. It is suggested that the standard viewing window be used when first graphing, and then modifications can be made as appropriate.

6. 3(*x* + 2) - 9 = 15

   x = ______

   WINDOW values:

   XMin: _______ XMax: _______ YMin: _______ YMax: _______

7. 3(*x* - 4) + 7 = 2(*x* + 4)

   x = ______

   WINDOW values:

   XMin: _______ XMax: _______ YMin: _______ YMax: _______

8. $\dfrac{x}{2}+\dfrac{x}{2}=20$

X = ______

WINDOW values:

XMin: _______ XMax: _______ YMin: _______ YMax: _______

NOTE: When using the graph screen to solve equations/inequalities, you should be aware that the displayed coordinate values approximate the actual mathematical coordinates. The accuracy of the displayed values is determined by the height and width of the pixel space being displayed. The space height/width formulas are discussed in detail in the Unit entitled "Preparing to Graph: Viewing Windows".

Identities and Contradictions

Algebraically	**Graphically**	**Numerically**	**Notation**
Solve $4(x - 1) = 4x - 4$ $4x - 4 = 4x - 4$ $-4 = -4$ Simplifying each side of the equation algebraically justifies that the equation is an **identity**. Identities are true for all values of *x* that are acceptable replacement values for the variable. Thus the solution to this equation is the set of all real numbers.	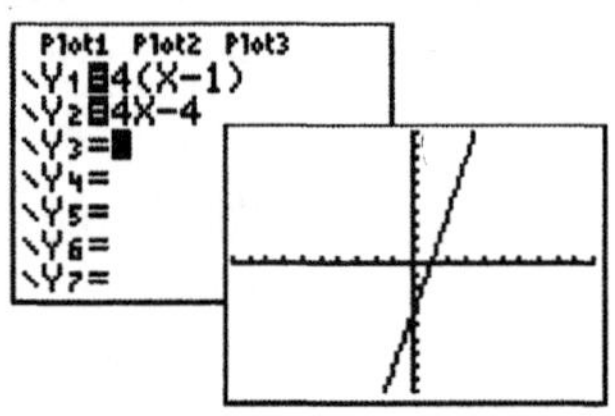 Press **[TRACE]** and trace along this line, observing the equation in the upper left corner of the screen. Use the up (or down) arrow key (press only one) and move the TRACE cursor to the other graph. Again check the equation displayed in the upper left corner or the number in the upper right corner, depending on the calculator. Which graph are you on now? Both graphs are the same! When both graphs are the same, then both sides of the equation must be equivalent expressions.	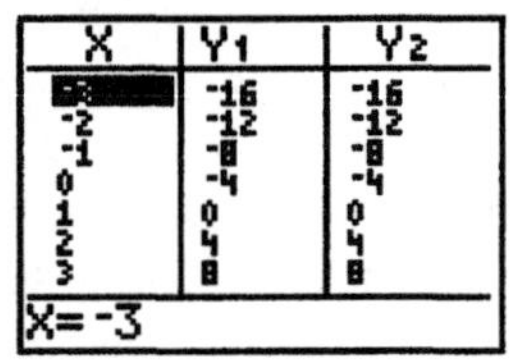 Look at the table of values. Both the left and right side of the equation have been evaluated for multiple values of *x*. Clearly, both sides of the equation are equivalent for each value of x. Equivalent expressions produce identical values for all replacement values of the variable.	Solution: The set of all real numbers. This is often written as $\mathbb{R}$.
Solve $2x - 5 = 2(x + 1)$ $2x - 5 = 2x + 2$ $-5 = 2$ After simplifying the original equation it is apparent that there is no solution. This equation is called a **contradiction** and has no solution.	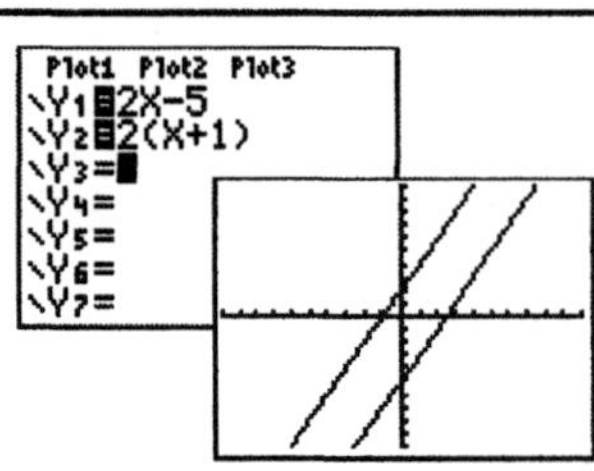 Observe that the two lines appear to be parallel.	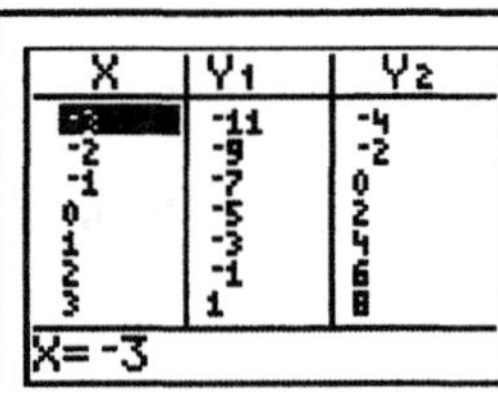 Notice that the values in the Y1 and Y2 column are 7 units apart for each displayed *x*-value.	Solution: No solution. This is often written as $\emptyset$.

EXERCISE SET CONT'D

9. Graphical representations of the three types of equations are displayed below. In the blank provided, identify each type of equation.

a. 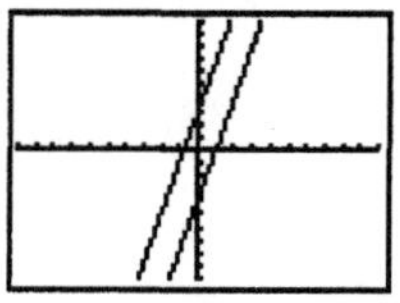

b. 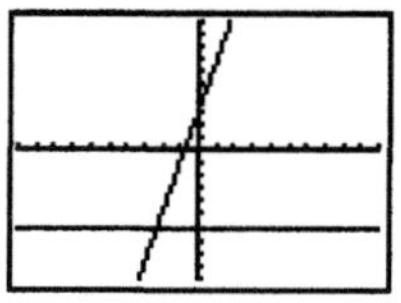

c. 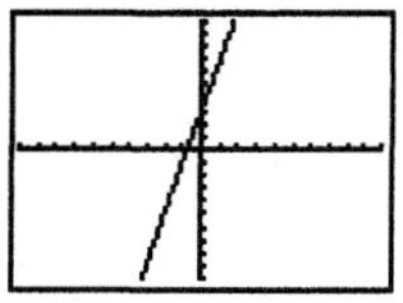

a. ______________ b. ______________ c. ______________

10. The pictured table represents one of the three types of equations. Determine the type of equation *and* the solution to the equation.

type:______________ solution:____________

X	Y_1	Y_2
-2	-5	11
-1	-1	11
0	3	11
1	7	11
2	11	11
3	15	11
4	19	11

X= -2

11. The pictured table represents one of the three types of equations. Determine the type of equation *and* the solution to the equation.

type:______________ solution:____________

X	Y_1	Y_2
-2	-6	-9
-1	-2	-5
0	2	-1
1	6	3
2	10	7
3	14	11
4	18	15

X= -2

12. The pictured table represents one of the three types of equations. Determine the type of equation *and* the solution to the equation.

type:______________ solution:____________

X	Y_1	Y_2
-2	-11	-11
-1	-8	-8
0	-5	-5
1	-2	-2
2	1	1
3	4	4
4	7	7

X= -2

13. Troubleshooting: From the pcitured displays a student concluded that the equation was a contradiction and the solution was ∅.

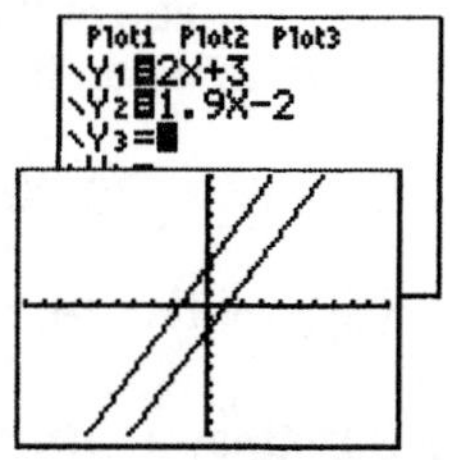

a. Was it reasonable for him to base his conclusion on the graphical display alone? Why or why not?

b. Which of the displays best confirms (for you) the fact that the equation was *not* a contradiction?

X	Y_1	Y_2
-2	-1	-5.8
-1	1	-3.9
0	3	-2
1	5	-.1
2	7	1.8
3	9	3.7
4	11	5.6

X= -2

<u>Solutions to Exercise Sets:</u> 1. -.4285714, $-\frac{3}{7}$ **2.** 1.5, $\frac{3}{2}$ **3.** 5.25, $\frac{21}{4}$ **4.** -.075, $-\frac{3}{40}$

5. The X-value is the solution to the equation. The Y-value is the quantity each side of the equation will evaluate to when the variable is replaced with the solution value.

6. 6; Xmin = -10, Xmax = 10, Ymin = -10, Ymax = at least 16

7. 13; Xmin = -10, Xmax = at least 14, Ymin = -10, Ymax = at least 35

8. 20; Xmin = -10, Xmax = at least 21, Ymin = -10, Ymax = at least 21

9a. contradiction **9b.** conditional **9c.** identity

10. Type: conditional; Solution: 2

11. Type: contradiction; Solution: $\varnothing$

12. Type: identity; Solution: $\mathbb{R}$

13. a. Based on the graphical display, it was a reasonable conclusion. **b.** The TABLE confirms that it is not a contradiction because the difference in values of the Y1 and Y2 columns is not constant. *Note: If you understand the concept of slope, the y-edit screen confirms the fact that the lines are NOT parallel because the slopes are different.*

Verbally	Algebraically	Numerically	Graphically	Notation
Find the solution(s) to the equation $5x - 1 = -3$.	$5x-1=-3$ $5x=-2$ $x=-\frac{2}{5}$	$5\left(-\frac{2}{5}\right)-1=-3$ $-2-1=-3$ $-3=-3$ 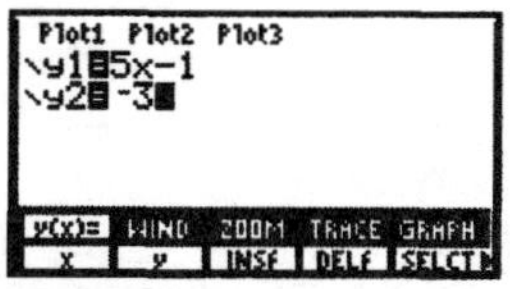  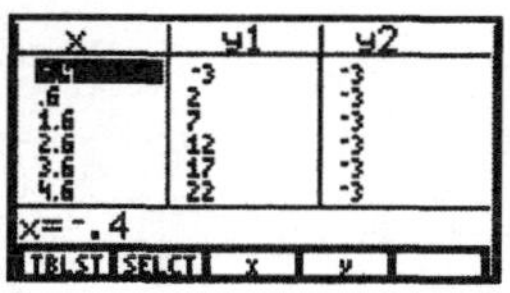 This is a verification of the algebraic solution.	 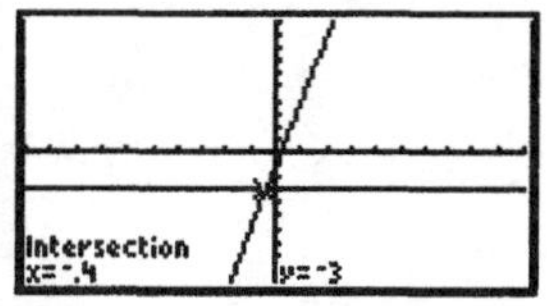 **Keystrokes:** • access INTERSECT option: **[2nd] <M5> (GRAPH).** •**[MORE] [F1] (MATH) [MORE] [F3] (ISECT)** • Move the cursor along the first curve to the approximate point of intersection. • **[ENTER]** • At the *second curve* prompt, **[ENTER]**. • At the *guess* prompt, press **[ENTER]**. You do not have to guess because there is only one point of intersection. The solution is -0.4.	• The solution, as read from the graphical display, is $x = -.4$. • The number - 0.4 is equivalent to the fraction, $-\frac{2}{5}$. To convert the decimal to a fraction, from the INTERSECTION screen, return to the home screen (**[2nd] <QUIT>**) and convert the value stored at x to a fraction (**[xVAR] [CUSTOM]** the appropriate F key for ▸Frac **[ENTER]**). 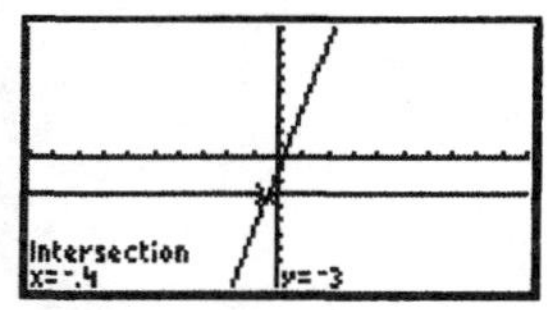 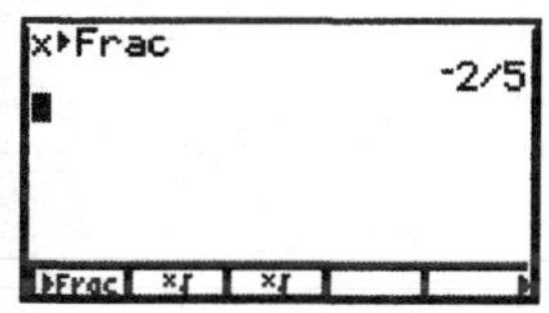 • Pressing **[EXIT]** or **[CLEAR]** removes the menus displayed at the bottom of the graph.

☞ RETURN TO THE EXERCISES ON PAGE 38 .

UNIT 7
GRAPHICAL SOLUTIONS: ABSOLUTE VALUE EQUATIONS

*Prerequisite: Unit #6

In the prerequisite unit, linear equations are solved graphically using the INTERSECT feature of the calculator. This same approach can be used to solve equations containing absolute value. Remember, absolute value is located by pressing **[MATH] [▸]** for the **NUM** menu or by pressing **[2nd] <CATALOG>**. (TI86 users should have this feature customized).

Algebraically	**Numerically**	**Graphically**	**Notes**
$\lvert x+3\rvert = 6$ $x + 3 = 6$ or $x + 3 = -6$ $x = 3$ or $x = -9$	$\lvert x+3\rvert = 6$ $\lvert 3+3\rvert = 6$ $\quad$ $\lvert -9+3\rvert = 6$ $\lvert 6\rvert = 6$ $\quad$ $\lvert -6\rvert = 6$ $6 = 6$ $\quad$ $6 = 6$ 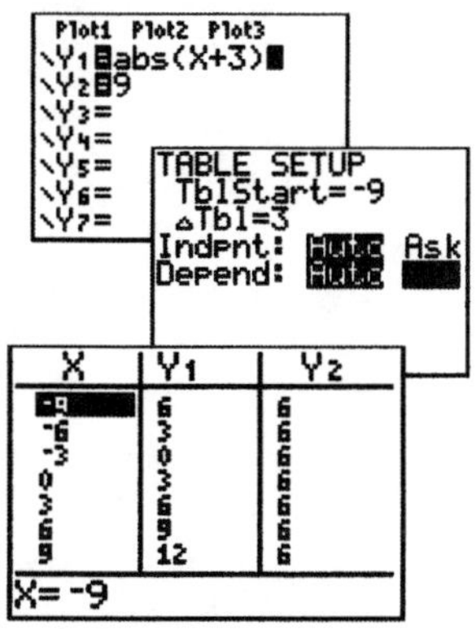 Note: In the TABLE SETUP, notice that the TABLE was incremented by units of 3 so that both solutions could be displayed in the *x* column without scrolling.	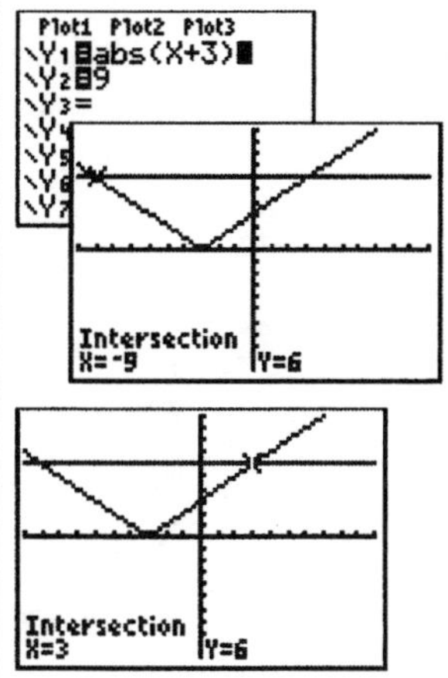 • Remember, always begin graphing in the standard viewing window (Zstandard). • Because there are two points of intersection, the INTERSECT feature will have to be completed two times. • At the GUESS prompt move the cursor close to the desired point of intersection.	In the future there may be other equations that have more than one solution. There will be an intersection point for each of these solutions that is a real number. The INTERSECT option will have to be completed for each point of intersection. To justify your work you will only be required to sketch one of the INTERSECT screens that you used and merely record the solutions derived from the other screens. The solutions, 3 and - 9, have been confirmed analytically, by substitution, and through the use of the TABLE and INTERSECT features of the graphing calculator.

EXERCISE SET

Directions: Graphically solve each of the equations below. Sketch the screen. Circle the points of intersection. Use the INTERSECT option to find the intersections. REMEMBER: Because there are two points of intersection, the process will have to be done twice. Record both of the solutions on the blanks provided.

1. $\lvert 2x - 1\rvert = 5$

   $x =$ ____ or $x =$ ____

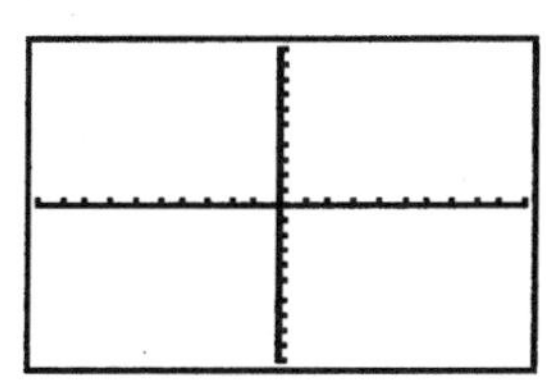

2. $\left|\frac{1}{2}x+1\right|=3$

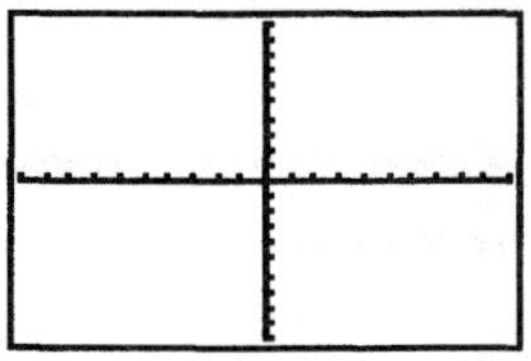

$x =$ ____ or $x =$ ____

3. $\left|\frac{4-x}{2}\right|=\frac{8}{5}$

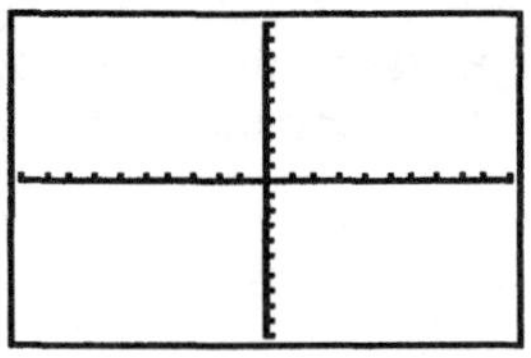

$x =$ ____ or $x =$ ____

4. $|x + 1| = |2x - 3|$

$x =$ ____ or $x =$ ____

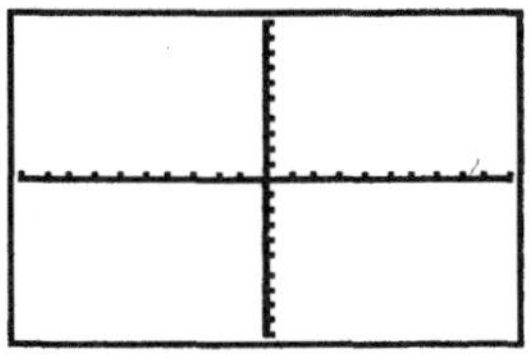

5. $|2 - x| = |3x + 4|$

$x =$ ____ or $x =$ ____

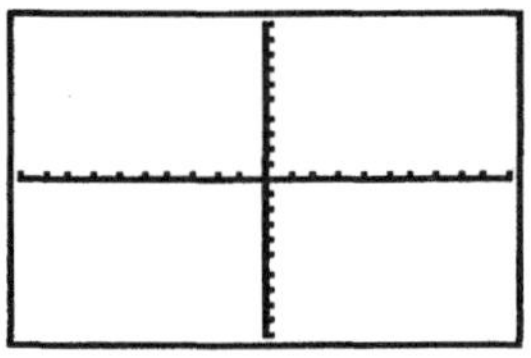

6. $|4x + 5| = -2.$

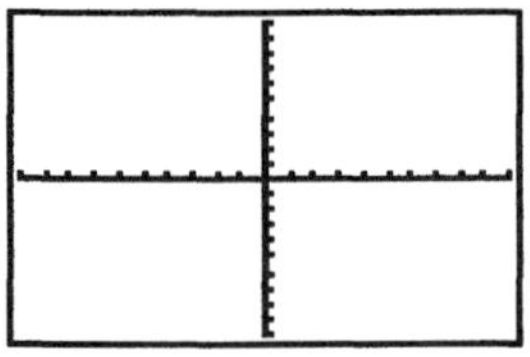

Do the graphs intersect?____

What is the solution? _____

✍ 7. The solution of the equation $|4x + 5| = -2$ (#6) can be determined by merely <u>looking</u> at the problem. What clue is there for this?

Directions: Graphically solve each of the equations below. Sketch the screen (axes will need to be adjusted and WINDOW values recorded in the spaces provided). Circle the points of intersection. Use the INTERSECT option to find the intersections. REMEMBER: Because there are two points of intersection, the process will have to be done twice. Record both of the solutions on the blanks provided.

8. $|2(x - 5) - 9| = 5$

Xmin= _____ , Xmax = _____, Ymin=______, Ymax= _____

$x =$ _______ or $x =$ _______

9. $|2x + 7| = 11$

Xmin= ____ , Xmax = ____, Ymin=____, Ymax= ____

x = ______ or x = ______

10. $|4x - 3| = |2x + 5|$

Xmin= ____ , Xmax = ____, Ymin=____, Ymax= ____

x = ______ or x = ______

11. $|3x - 1| = |7 + 4x|$

Xmin= ____ , Xmax = ____, Ymin=____, Ymax= ____

x = ______ or x = ______

12. $|x - 3| = -|x + 20|$

Xmin= ____ , Xmax = ____, Ymin=____, Ymax= ____

Do the graphs intersect?______

What is the solution?______

Hint: Try zooming out by pressing **[ZOOM] [3] (3:Zoom Out) [ENTER]**. TI-86 users press **[GRAPH] [F3] (ZOOM) [F3] (ZOUT) [ENTER]**.

Solutions:

Exercise Set: 1. x= -2 or x=3 **2.** x=-8 or x=4 **3.** x=0.8 or x=7.2 **4**. x= 2/3 or x= 4

5. x= -3 or x=-0.5 **6.** No, null set **7.** An absolute value cannot be set equal to a negative number.

8. x = 7 or x = 12, Xmax should be larger than 12 **9.** x = -9 or x = 2, Ymax at least 12

10. x =-1/3 or x = 4, Ymax at least 14 **11.** x = -8 or x = -6/7, Ymax at least 26

12.. Xmin = -30, no, ϕ,

UNIT 8
GRAPHICAL SOLUTIONS: QUADRATIC AND HIGHER DEGREE EQUATIONS

*Prerequisite: Unit #6

This unit considers the ROOTor ZERO method as another graphic approach to solving equations. This is in addition to the INTERSECT method (introduced in a prior unit). The ROOT/ ZERO method can be used to find the REAL roots/zeroes of all the equations presented thus far and for any equation encountered in the future that can be solved graphically. It is possible to divide the unit into two parts if the accompanying textbook addresses quadratic equations in a section separate from higher order equations.

FACTORABLE EQUATIONS

Algebraically

Solve: $x^2 - 2x = 8$

$x^2 - 2x - 8 = 0$

$(x - 4)(x + 2) = 0$

$x - 4 = 0$ or $x + 2 = 0$

$x = 4$ or $x = -2$

Numerically

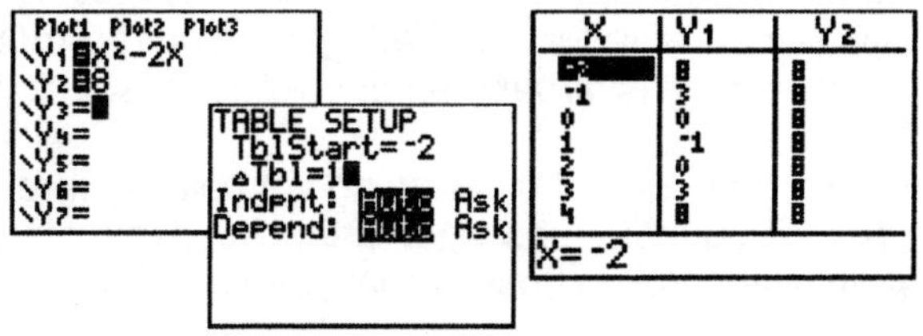

Graphically/Intersect

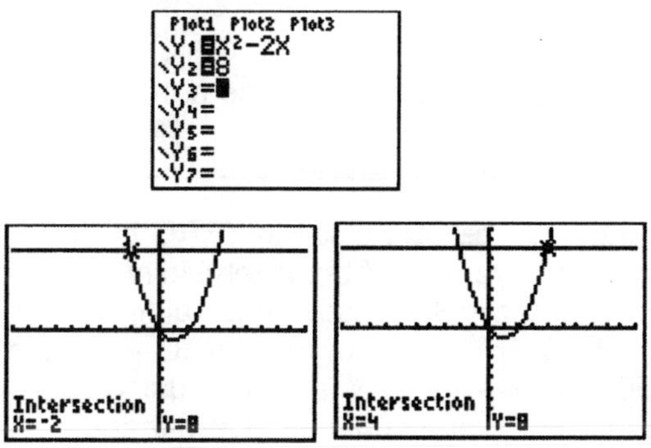

INTERSECT option: press **[2nd] <CALC> [5] (5:Intersect)** and respond to the prompts. Because there are two points of intersection, the INTERSECT feature must be used twice. At the GUESS prompt, cursor near the desired point of intersection.

TI-86 Users: If more than one line of menus are displayed, press **[EXIT]** to ensure only one line of menu options is displayed. Press **[MORE] [MATH] [MORE]** and the appropriate F key to access the ISECT option.

Graphically/ROOT/ZERO

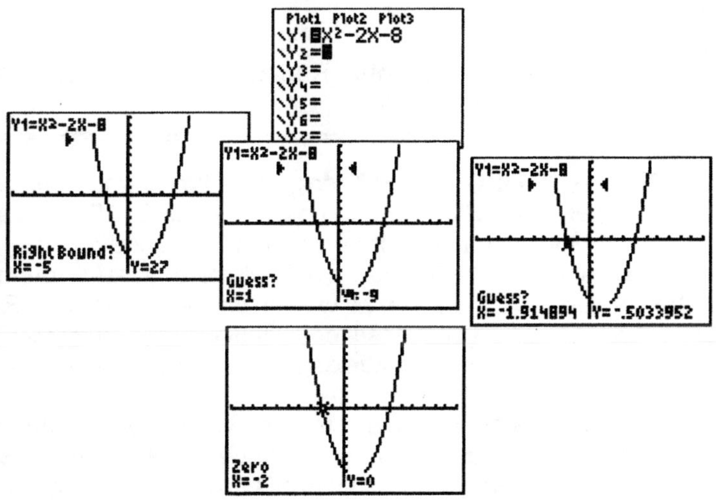

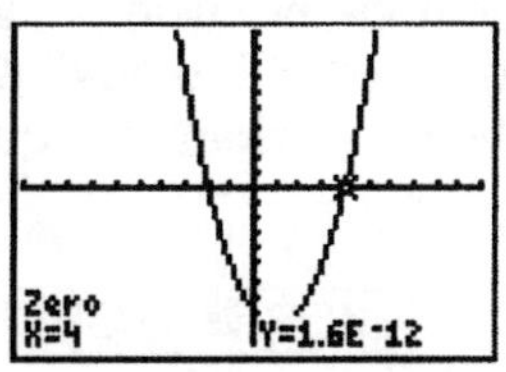

Bounds were set for this root in the same manner as for the left hand root.

To access the ROOT/ZERO feature, press **[2nd] <CALC> [2] (2:zero)** .

TI-86 Users: Before accessing the ROOT option, be sure that only one line of menu options is displayed. If more than one line is displayed, press **[EXIT]**. Now, press **[MORE] [F1](MATH)** and press the appropriate F key to select **(ROOT)**.

Set bounds as specified in the boxed information below this box.

STEPS FOR SOLVING AN EQUATION USING THE ROOT/ZERO FEATURE

a. **Set left bound:** The screen display asks for a left bound. A left bound is an *x*-value smaller than the expected root; move the cursor to the left of the left-hand root and press **[ENTER]**. Because the roots are determined on the horizontal axis, a left bound is always determined by moving the cursor to the left of the root. Notice at the top of the screen a ▸ marker has been placed to designate the location of the left bound. You have the option of not moving the cursor to bound the root, but rather to enter an appropriate value for *x* at the "left bound" prompt. Care should be taken that the value entered for a left bound is *smaller* than the expected solution. Press **[ENTER]** after entering the value.

b. **Set right bound:** Similarly the right bound is always determined by moving the cursor to the right of the root. At the right bound prompt, move the cursor to an x- value larger than the expected root and press **[ENTER].** Again, a ◂ marker is at the top of the screen to designate the location of the bound. Again, you have the option of not moving the cursor to bound the root, but rather to enter an appropriate value for *x* at the "right bound" prompt. Care should be taken that the value entered for an right bound is *larger* than the expected solution. Press **[ENTER]** after entering the value.

NOTE: If the bound markers do not point toward each other, ▸ ◂, then you will get an "ERROR:bounds" message. If this happens, start the ROOT/ZERO calculation over. Care must be taken when setting bounds that you bound *only* the specific root for which you are searching.

c. **Locate first root:** Move the cursor to the approximate location where the graph crosses the *x*-axis for your guess. When you press **[ENTER]** the calculator will search for the root, within the area marked by ▸ and ◂. The root is $x = 0$. Make sure you read the "Trouble Shooting Note" below. It addresses round-off error.

d. **Locate subsequent roots:** Repeat the entire process outlined above to determine the right-hand root. This root is $x = 1$.

TROUBLE SHOOTING NOTE: There will be times when the calculator will be very close to ZERO but will not display ZERO exactly. For example X = -7.65 E -15 is the calculator's version of the scientific notation -7.65×10^{-15} which is equivalent to X =- .00000000000000765. For all practical purposes this value is ZERO. To verify that the X value is actually zero, scroll through the TABLE to $x = 0$ (or use EVAL) and note that the *y*-value is -4, the same as was displayed on the INTERSECT screen. Therefore, when using the graph screen to solve equations/inequalities with the ROOT/ZERO feature, you should be aware that the display coordinate values sometimes approximate the actual mathematical coordinates.

If an equation is not set equal to zero, then you should try the INTERSECT method first. WHY? By merely entering the existing equation into the calculator (left side at the Y1 prompt and right side at the Y2 prompt) you run less risk of error because there is no algebraic manipulation.
If an equation is set equal to zero, when the non-zero side of the equation is entered at the Y1 prompt and graphed, the real roots or zeroes are the *x*-values at the point(s) where the graph crosses the horizontal axis (*x*-axis). Therefore, the ROOT/ZERO method is **perfect** for equations of any type that are already set equal to zero.

EXERCISE SET

Directions: Solve each of the following quadratics, using the ROOT/ZERO option. Display the graph in the standard viewing window. Sketch your graph display, circle the two roots and record their values in the blanks provided. Beneath each problem, factor the quadratic that has been graphed.

1. $x^2 + 8x - 9 = 0$

   The roots are $x =$ _____ and $x =$ _____.

   Factorization:________________________________

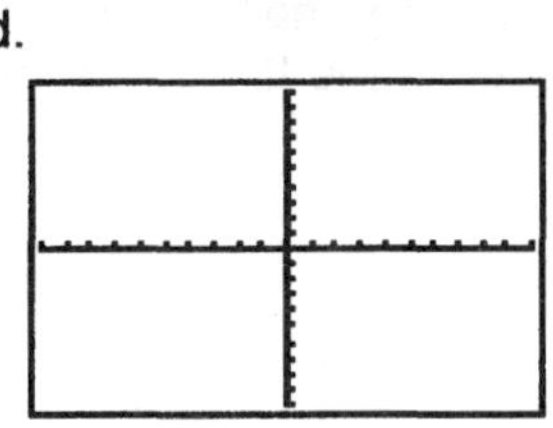

2. $(x - 2)^2 + 3x - 10 = 0$

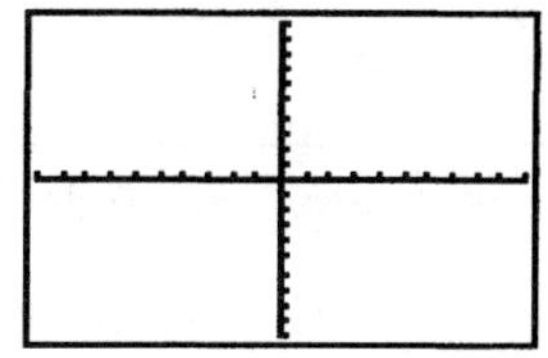

The roots are x = _____ and x = _____.

Factorization:______________________________

3. $x^2 + 6x + 9 = 0$

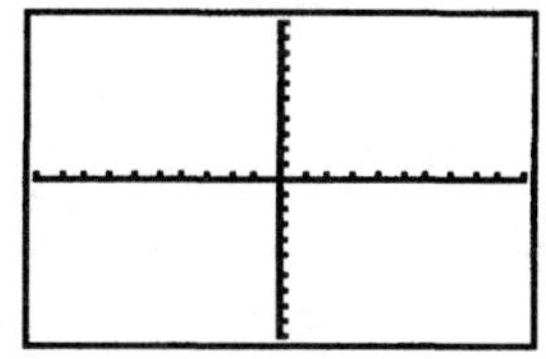

The root(s) is/are x = _____ .

Factorization:______________________________

(Note: the point at which the graph touches, but does not cross the x-axis, produces two identical roots - often called a double root.)

✍4. Compare the factorization of the polynomial to the real roots that were determined in each of the problems above. How do they compare?

5. If you know that the roots to an equation are 4 and -2, you should be able to write an equation, in factored form that has these roots. Write an equation, in factored form:

____

6. The equations that have been solved thus far in this unit have all been second degree equations. Based on the exercises, how many roots should you expect to have with a second degree (quadratic equation)? ________________

✍ 7. **ONE** of the equations does not conform to the pattern. Which one is it and why is the number of roots different from the rest of the problems?

Directions: The next set of equations contain polynomials that are not second degree. However, these polynomials can be factored.

a. First use the ROOT/ZERO method to solve the equation.
b. Copy your screen display.
c. Circle the real roots.
d. Record the value of the roots in fractional form.
e. Factor the polynomial.

8. $x^3 - 7x^2 + 10x = 0$

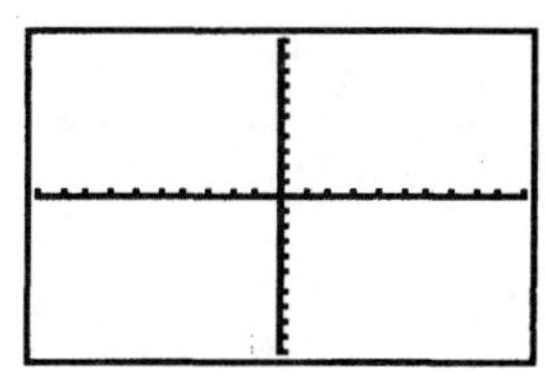

The roots are x = _____, _____ and _____.

Factorization:________________________

9. $x^4 - 5x^2 + 4 = 0$

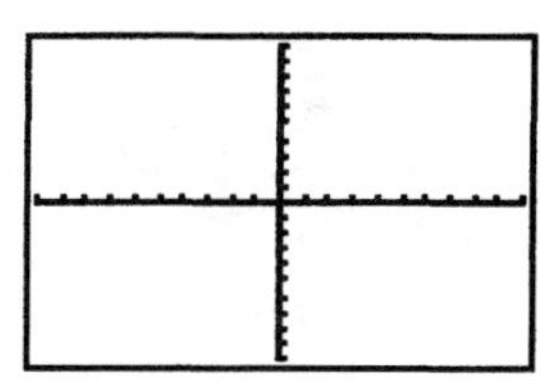

The roots are x = _____, _____, _____ and _____.

Factorization:________________________

10. $3x^4 + 2x^3 - 5x^2 = 0$

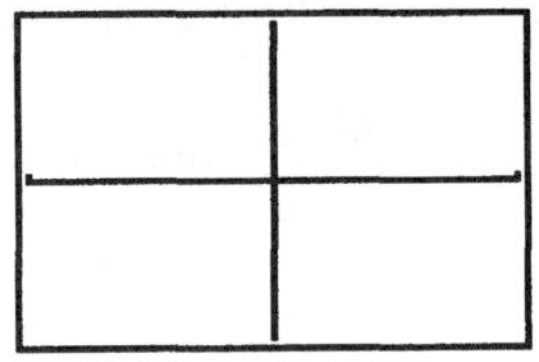

Suggestion: Change the WINDOW values (Xmin= -5, Xmax= 5) OR use the **ZBox** option.

The roots are x = _____, _____ and _____.

Note: There is a double root in this problem. Which root is the double root?____

Factorization:____________________

✍11. What conclusions can be drawn about the number of real solutions and the degree of the polynomial equations that have been solved thus far?

NON-FACTORABLE EQUATIONS

Previously, factorable polynomial equations were solved. Several observations were made: 1) for each factor that contained a variable there was a root; 2) the number of factors containing a variable corresponded to the degree of the polynomial; 3) the number of real roots was equal to or less than the degree of the polynomial. CONCLUSION: A polynomial equation of degree n will have <u>at most *n* real roots</u>. We will now examine polynomial equations that do not factor over the rational numbers and hence may not have any real roots, or at best roots that are irrational.

Example 4: Solve the equation $3x^2 - 18x + 25 = 0$.

Algebraically	**Graphically**
Because the polynomial cannot be factored we use the quadratic formula to solve: $3x^2 - 18x + 25 = 0$ $a = 3,\quad b = -18,\quad c = 25$ $\frac{-b \pm \sqrt{b^2 - 4ac}}{2a} =$ $\frac{-(-18) \pm \sqrt{(-18)^2 - 4(3)(25)}}{2(3)} =$ $\frac{18 \pm \sqrt{324 - 300}}{6} =$ $\frac{18 \pm \sqrt{24}}{6} =$ $\frac{18 \pm 2\sqrt{6}}{6} = \frac{9 \pm \sqrt{6}}{3} = 3 \pm \frac{\sqrt{6}}{3}$ These root/solutions are exact values.	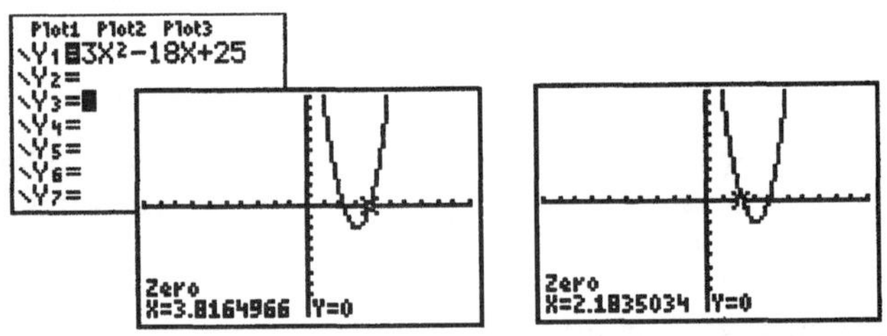 Enter $3x^2$ - 18x + 25 after Y1= and graph in the standard viewing window. Use the ROOT/ZERO option twice to compute both roots of the equation. The screens displayed indicate both roots. These roots should be comparable to the approximate values that were determined at the left. This is displayed on the screen below. 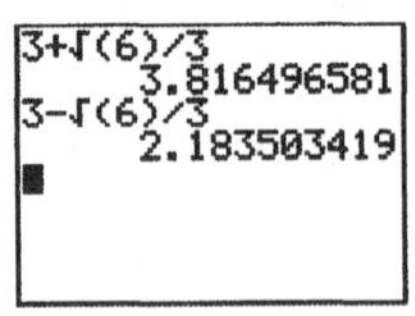 Irrational roots are displayed as approximations on the calculator.

EXERCISE SET CONT'D

Directions: Use the ROOT/ZERO option to find the **REAL** roots of the following quadratic equations. Sketch the screen display in the indicated viewing window and record the solutions.

12. $-5x^2 + 5x + 8 = 0$
    $x =$ ______________ and ______________

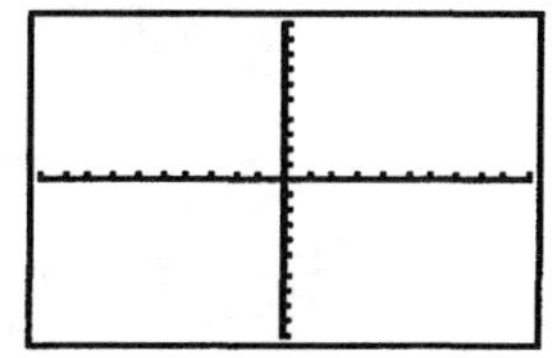

ZStandard

✍13. What happens when you try to convert the roots in #12 to fractions? Why?

14. Solve the quadratic equation in #12 by either completing the square or using the Quadratic Formula. From the home screen, approximate the solutions and compare them to the calculator answers recorded. They should be the same.

✍15. $x^2 + 5x + 8 = 0$

Is it possible to graphically find the roots of this equation using the calculator?________

Why not?

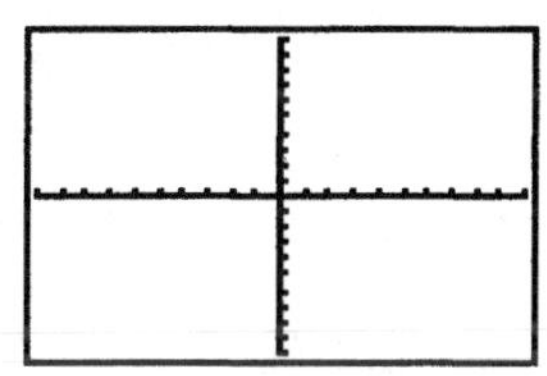

16. Solve #15 analytically using either the Quadratic Formula or by completing the square.

Graphs which intersect the x-axis will have real roots because the x-axis represents the set of real numbers. Roots which are complex numbers will not be represented on the x-axis. Thus an equation with complex roots will not intersect the x-axis. There will, however, be two roots because complex roots always occur in conjugate pairs.

17. State the number and type of roots (real or complex) of each of the following. DO NOT SOLVE, simply graph the polynomial function in the ZStandard viewing window and check the number of x-intercepts, if any.

a. $6x^2 + 2x - 4 = 0$

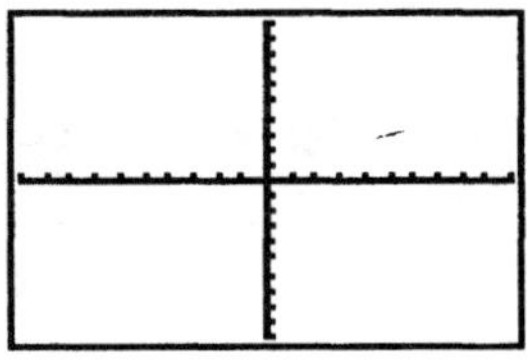

Number of roots:_______

Type of roots:_________

b. $2x^3 - 5x + 5 = 0$

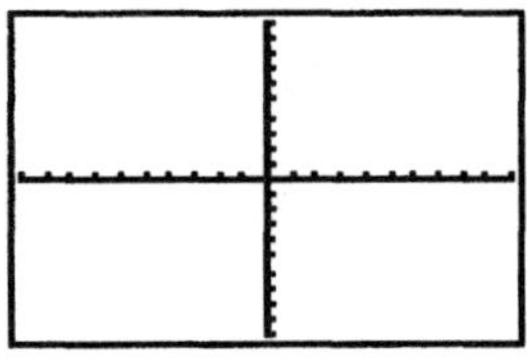

Number of roots:_______

Type of roots:_________

c. $x^4 - 0.5x^3 - 5x^2 + 10 = 0$

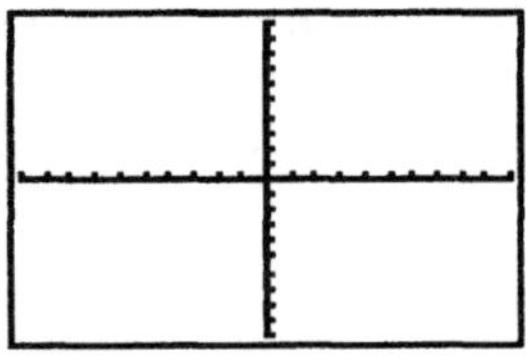

Number of roots:_______

Type of roots:_________

18. $0 = x^4 - 2x^3 + x - 2$

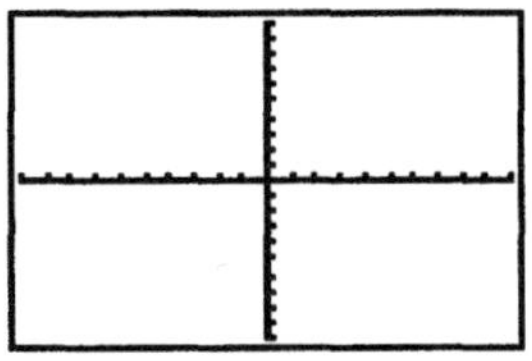

x = _____________ and _____________

The expression $x^4 - 2x^3 + x - 2$ in completely factored form is $(x - 2)(x + 1)(x^2 - x + 1)$. Use the appropriate algebraic technique to determine the complete solution set.

____

Solutions: **1**. $(x+9)(x-1)=0$, $x=-9$ and $x=1$ **2.** $(x-3)(x+2)=0$, $x=-2$ and $x=3$,

3. $(x+3)(x+3)=0$, $x=-3$ **4.** The factor is always the "variable minus the root".

5. $(x-4)(x+2)=0$ **6.** two **7.** #3. We did not see two distinct real roots, but rather two identical roots (often called a double root).

8. $x(x-5)(x-2)=0$, $x=0,2$ and 5 **9.** $(x-2)(x+2)(x-1)(x+1)=0$, $x=2,-2,1$ and -1

10. $x^2(3x+5)(x-1)=0$, $x=0,-5/3$ and 1, Zero is the double root. **11.** The number of solutions is the same as the degree of the equation. However, not all the solutions are distinct (different). Some appear as multiple roots. **12**. $x=-.860147$ and $x=1.8601471$

13. These answers will not convert to fractions because they are the decimal approximations of the irrational numbers $\frac{1}{2} \pm \frac{\sqrt{185}}{10}$. **14.** $\frac{5 \pm \sqrt{185}}{10}$ **15.** No, It does not intersect the x-axis and therefore has no real roots.

16. $x = \frac{-5 \pm i\sqrt{7}}{2}$ **17. a.** 2, real **b**. 3, 1 real, 2 complex **c.** 4, complex **18.** $x = 2$, $x = -1$, $\left\{2, -1, \frac{1 \pm i\sqrt{3}}{2}\right\}$

UNIT 9
GRAPHICAL SOLUTIONS: RADICAL EQUATIONS

*Prerequisite: Unit #8 and #4

This unit uses the INTERSECT and ROOT/ZERO options on the calculator to solve equations that contain radicals. Recall, the solution to an equation is the value(s) for the variable that produce a true arithmetic statement.

<table>
<tr><th>Algebraically</th><th>Numerically</th><th>Graphically</th></tr>
<tr>
<td>$\sqrt{3x+7}+2=7$
$\sqrt{3x+7}=5$
$3x+7=25$
$3x=18$
$x=6$</td>
<td>$\sqrt{3(6)+7}+2=7$
$\sqrt{25}+2=7$
$5+2=7$
$7=7$
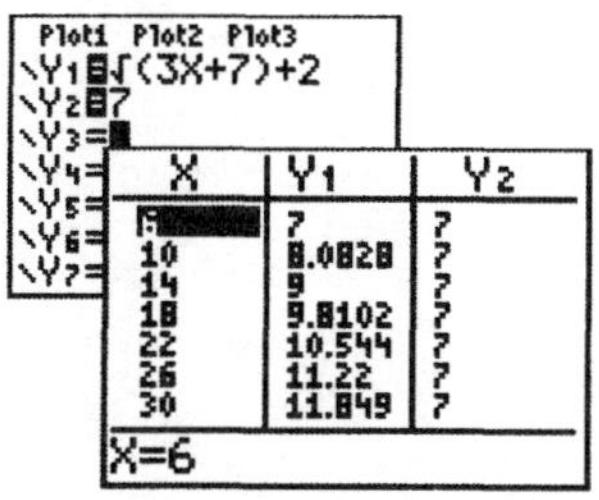
</td>
<td>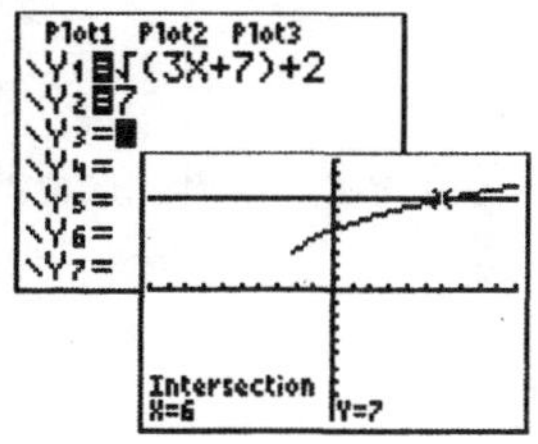

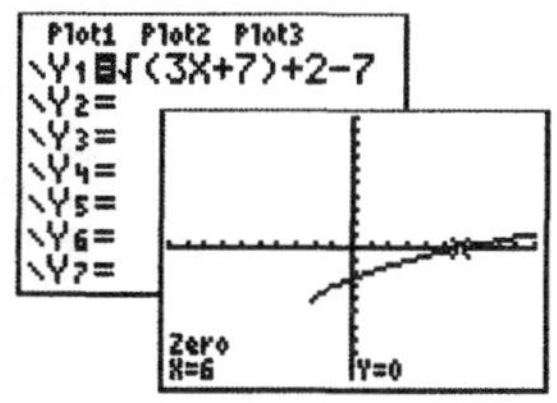

To use the intersect feature to solve graphically, make sure an expression is entered at both the Y1 and the Y2 prompts.

To use the ROOT/ZERO feature for solving, one side of the equation must be 0. You then enter the polynomial side at the Y1 prompt.

When responding to prompts make sure a value is displayed for both *x* and *y* on the graph screen before pressing ENTER to respond to the prompt.</td>
</tr>
</table>

EXERCISE SET

Directions: Use either the ROOT/ZERO or the INTERSECT options on the calculator to solve each radical equation below. Sketch the screen display and use the ▸Frac option (under the MATH menu) to convert all decimal results to fractions.

1. $\sqrt{x^2+6x+9}=-x+6$

   $x =$ ________

   Converted to a fraction, $x =$ ________

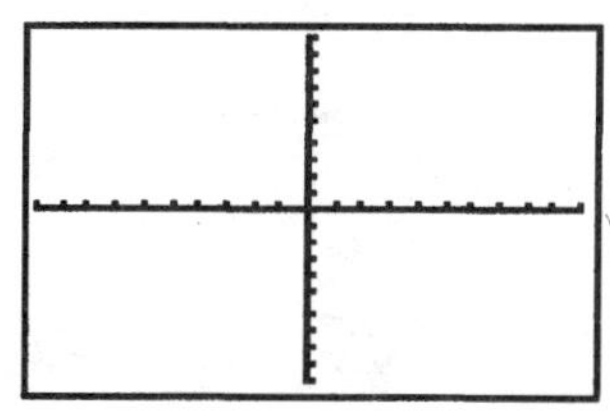

2. $\sqrt{2x+5}=\sqrt{3-x}$

   $x =$ ________

   Converted to a fraction, $x =$ ________

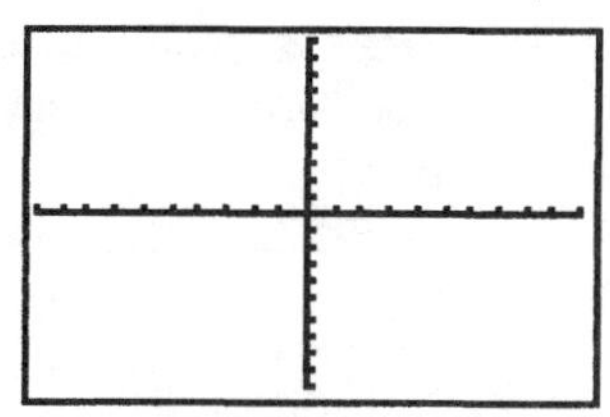

3. $\sqrt[3]{2x+6}=2$

   $x =$ ________

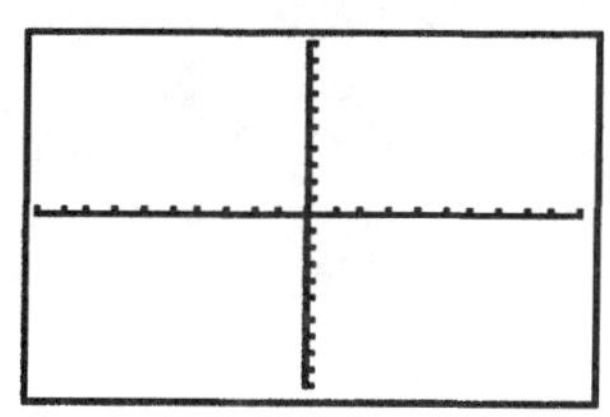

4. $\sqrt{x+4}+6=3$

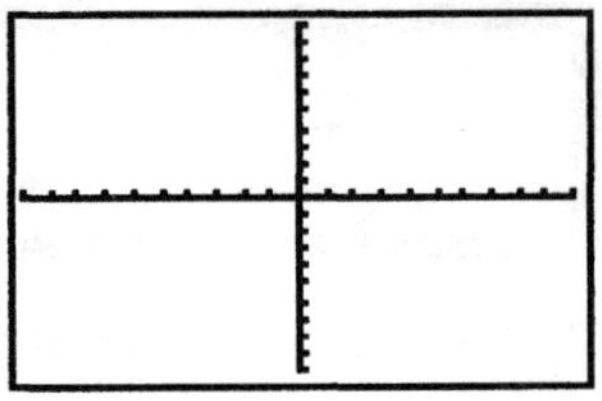

Solution: ____________

5. If you solved the equation in #4 algebraically, the first step would be to isolate the radical. Once the radical is isolated, you should realize there are no solutions. Why?

6. Solve the equation $\sqrt{x}+2=\sqrt{5-x}+3$ algebraically in the space below. Check your solution(s) by substitution.

Algebraic Solution

Check by substitution

7. Graph the equation in #6 by entering the left side at Y1 and the right side at Y2. Copy the display screen. How many points of intersection do you see? Use the calculator to find the solution.

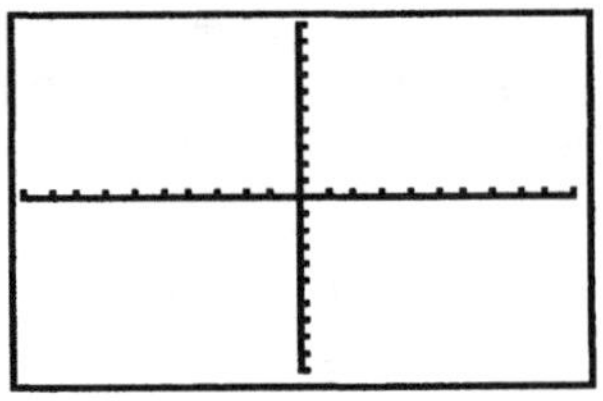

$x =$ __________

8. $\sqrt{2x-3}=3-x$

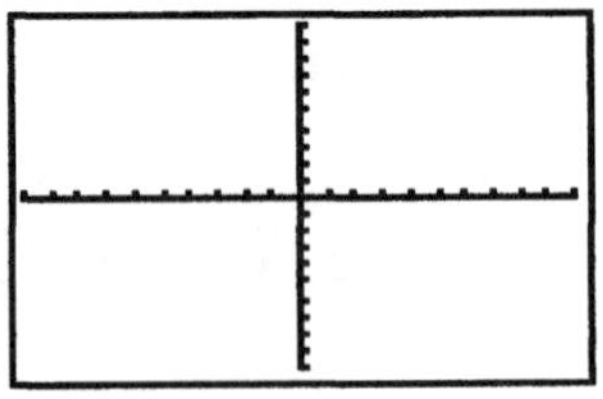

Be Careful! Make sure BOTH the x and y coordinates are displayed at the bottom of the screen.

$x =$ ________

9. $\sqrt{x^2-12x+36}+5=7$

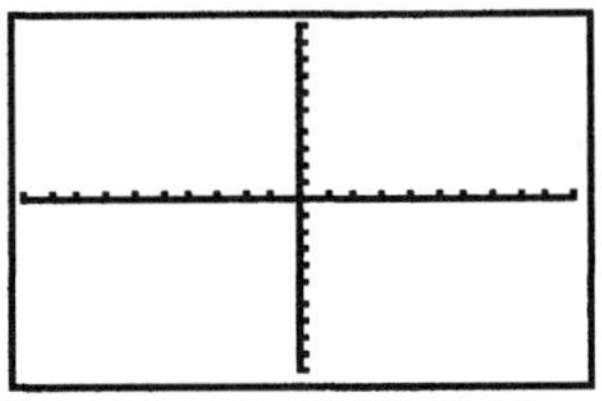

$x =$ ________ $x =$ ________

10. $\sqrt[3]{x^3+5x^2+9x+18}=x+2$

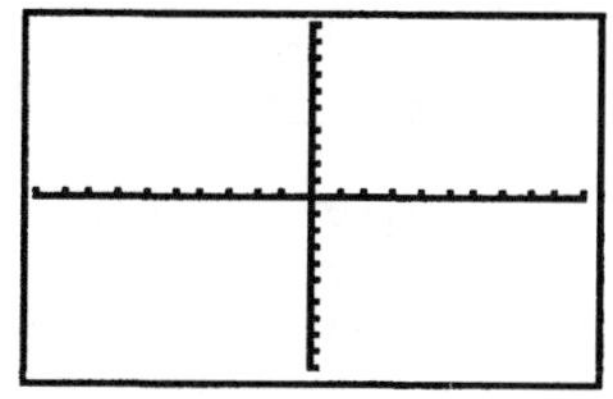

x = ________ x = ________

Hint: If you have difficulty finding the roots using the INTERSECT or ROOT/ZERO option, access the TABLE feature (set the table to begin with $x = 1$ and increment by 1). Enter the left side of the equation at Y1 and the right side at Y2 (as though you were using the INTERSECT option). Access the table, and scroll until you find the x-value(s) for which the Y1 and Y2 values are equal.

11. In the space below, solve $\sqrt{-2x+6}=3-x$ algebraically. Check solutions by substitution.

Algebraic solution: Check by substitution:

12. a. Use the INTERSECT option to find the solutions to the equation in #11. Copy your screen display.

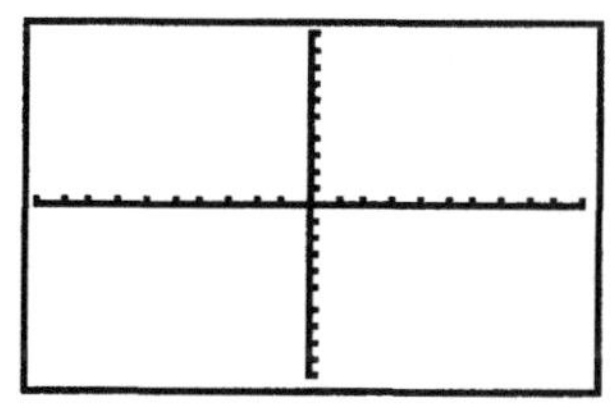

x = ________

b. Use the ROOT/ZERO option to find the solutions. Copy your screen display.

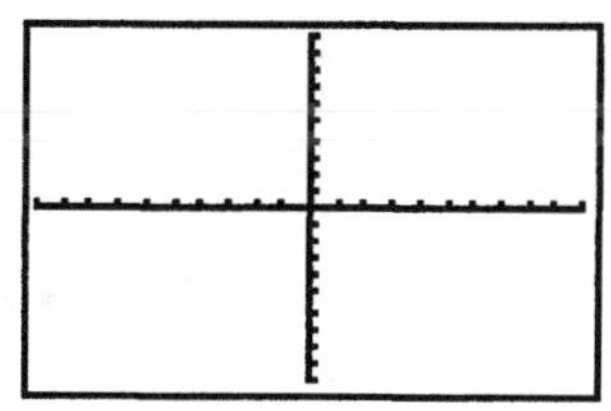

x = ________

c. You found two valid roots algebraically, two roots were displayed graphically, and yet only one root could be computed with the calculator. Which is the correct solution - the algebraic solution in #11 or the calculator solutions above?

d. Set your table to a start value of 1 and increment by 1. Make sure you have the left side of the equation entered at Y1 and the right side at Y2. Access the table. For what values of x are the Y1 and Y2 values equal?

x = ________ x = ________ This confirms your algebraic solution.

e. Explain **why** the calculator was unable to compute both roots in part a.

13. **Application:** The period of a pendulum on a clock is the time required for the pendulum to complete one cycle (one "swing" from a given position back to this initial point). The formula for finding the period of a pendulum is $T = 2\pi\sqrt{\frac{x}{32}}$, where T is the time required in seconds and x is the length of the pendulum. A clock company is constructing a clock for a window display. If it takes the pendulum two seconds to complete 1 period, what is the length of the pendulum (to the nearest hundredth of a foot)?

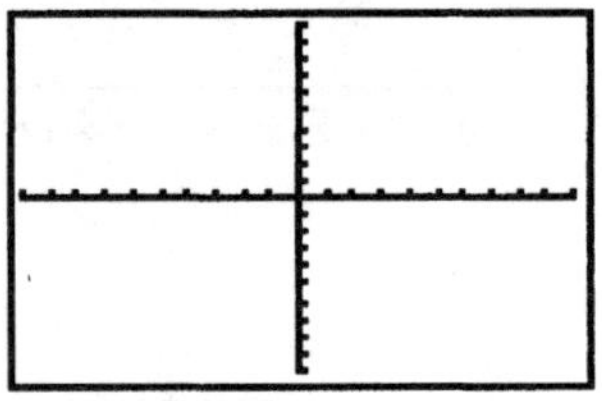

NOTE: Work problems from your text using what you have learned in this unit. Decide what works best for <u>you</u> - algebraic solutions? the ROOT/ZERO option? the INTERSECT option? Then PRACTICE.

<u>**Solutions:**</u> **1.**. $x = 1.5, 3/2$ **2.** $x = -.6666667, -2/3$ **3.** 1 **4.** Null Set

5. Because the right side would be equal to -3 and $\sqrt{\ }$ is defined only for positive roots.

6. $x = 4$, One is an extraneous root. **7.** $x = 4$ **8.** $x = 2$ **9.** $x = 4$ or $x = 8$

10. $x = -5$ or $x = 2$ **11.** $x = 1$ or $x = 3$

12a. $x = 1$ **12b.** $x = 1$ **12c.** {1, 3} - Both of these solutions check algebraically.

12d. $x = 1$ or $x = 3$

12e. When using the INTERSECT feature, the calculator establishes upper and lower bounds using the domain of the graphed functions. It then searches between these bounds for the point of intersection, **excluding the bounds in its search**.

13. 3.24 feet

UNIT 10
GRAPHICAL SOLUTIONS: LINEAR INEQUALITIES

*Prerequisite: Unit #6

To solve linear inequalities such as $5x - 1 \geq -3$, we want to find replacement values for x that will produce a true arithmetic sentence. The solution to a first degree inequality in one variable is typically an infinite set of numbers rather than a single number.

Algebraically	**Numerically**	**Verbally**
$5x - 1 \geq -3$ $5x \geq -2$ $x \geq -\frac{2}{5}$ Any number greater than or equal to $-\frac{2}{5}$ is a solution to the original inequality.	$5\left(-\frac{2}{5}\right) - 1 \geq -3$ $-2 - 1 \geq -3$ $-3 \geq -3$ The TABLE feature allows us to see the evaluation of the polynomial for many values of x. 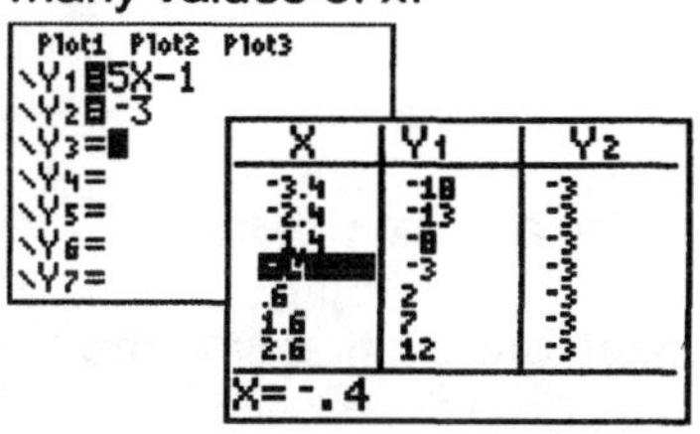	This is a true statement because -3 is equal to -3. Analytically, we can only check one number at a time. Note: -0.4 is the decimal equivalent of $-\frac{2}{5}$. When $x > -0.4$ the expression $5x - 1$ has values greater than - 3 and when $x < -0.4$, the expression $5x - 1$ has values less than negative 3.

Graphically	**Interpretation**	**Notation**
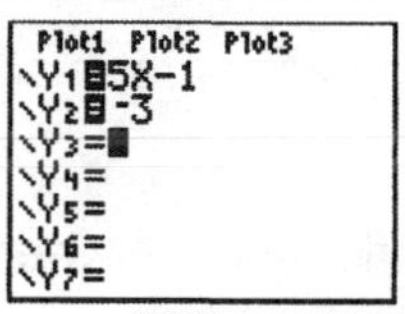 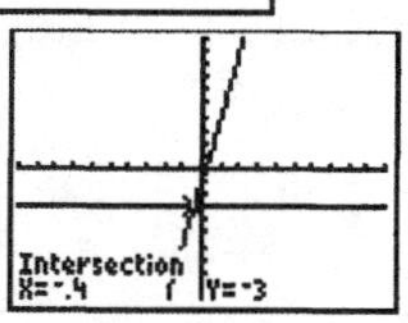 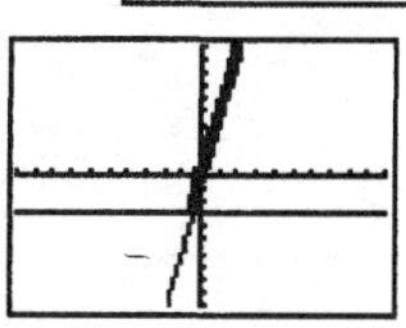	•Graphs should be read from left to right. The graph is a visual representation of numerical information that is displayed in the TABLE. • Use the INTERSECT feature to determine the critical point (the x-value for which Y1 = Y2). It is -3. •For what x-values is Y1 greater than Y2? It should be the section that is highlighted on the graph displayed. •In general, Y1 > Y2 where the graph of Y1 is <u>above</u> the graph of Y2.	The solution to the inequality is $x \geq -0.4$. In set notation this would be written: $\{x \mid x \geq -0.4\}$ As a number line graph this would be:  In interval notation this would be written $[-0.4, \infty)$.

The -2/5 you got in your solution is a "critical point." It is so named because it divides the number line into three distinct subsets of numbers - those larger than the number, those smaller than the number, and the number itself. The critical point is the solution of the *equation* associated with the given inequality. You can find the critical point by using the INTERSECT feature of the graphing calculator.

Steps for Graphically Solving a Linear Inequality

1. Enter the left side of the inequality at Y1 and the right side at Y2 on the y= (y-edit) screen.
2. Graph the equation in ZStandard (adjust window size if needed to see the point of intersection of the two lines) and sketch the display. Be observant so you can label the graphs as Y1 and Y2. The graph of Y1 will appear first.
3. Use the INTERSECT feature to find the point of intersection (the critical point). The expressions entered at Y1 and Y2 are equivalent at the intersection point.
4. Highlight the section of Y1 that is less than Y2 (for the inequality $Y1 < Y2$). That is where the graph of Y1 is below the graph of Y2. If your problem reads $Y1 > Y2$, you would highlight the portion of Y1 that is above the graph of Y2.
5. Determine the solution by projecting the shaded portion of the graph to the *x*-axis. These *x*-values are solutions to the inequality. Check the solution by setting TblStart equal to the critical point and examine the relationship between the values of Y1 and Y2 for *x*-values greater than and less than the critical point.
6. Draw your conclusions about the solution of the inequality.

EXERCISE SET

1. **Solve** $10 - 3X < 2X + 5$ graphically.

Solution Steps:

a. Press **[Y=]** and enter $10 - 3x$ after Y1= and $2x + 5$ after Y2=.

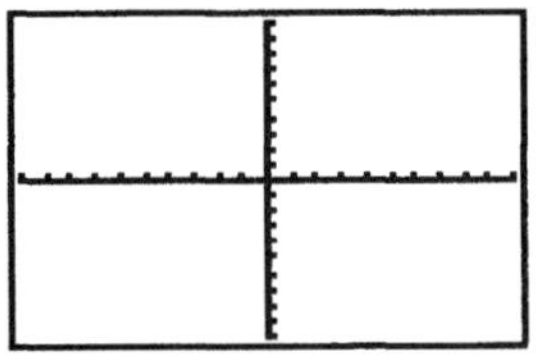

b. Sketch the graph displayed. Be observant the first time the graphs are displayed. Label Y1 and Y2. The graph of Y1 will appear first.

c. Use the INTERSECT feature to find the point of intersection. The intersection is $x =$ _____. The expressions entered at Y1 and Y2 are equivalent when $x = 1$. Therefore, $x = 1$ would be the solution to the <u>*equation*</u> $10 - 3x = 2x + 5$.

d. Highlight the section of Y1 that is less than Y2; where the graph of Y1 is <u>below</u> the graph of Y2.

e. Determine the solution by setting **TblStart**= *critical point*. Examine the relationship between the values of Y1 and Y2 for *x* values that are greater than, equal to, and less than the critical point.

f. Conclusion: $Y1 < Y2$ when $x > 1$.
The solution set will be $\{ x \mid x > 1\}$.
Translate this solution to a number line graph:

2. **Solve** $6 - 5x \leq -1$ graphically, following the steps outlined.

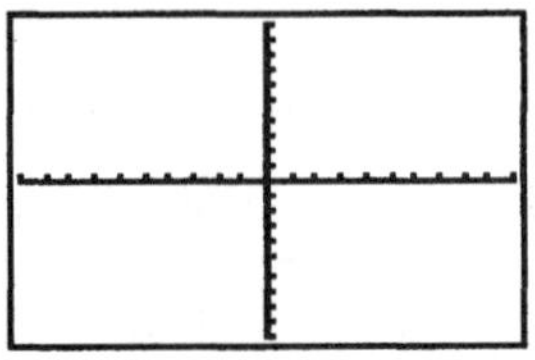

Solution Set:________________

Translate this solution to a number line graph:

3. **Solve** $-3 \geq 7 - 2x$ graphically, following the steps outlined.

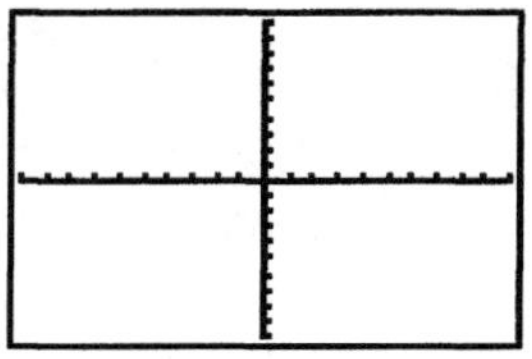

Solution Set:________________

Translate this solution to a number line graph:

4. **Solve** $\frac{4x-2}{6} < \frac{2(4-x)}{3}$ graphically, following the steps outlined.

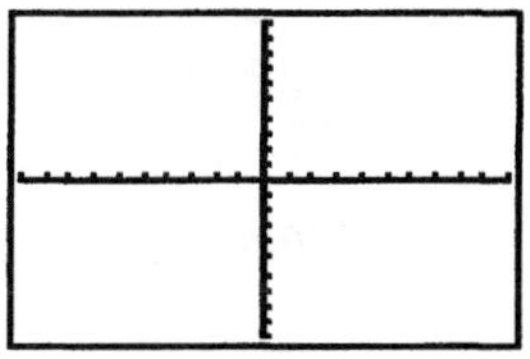

Solution Set:_____________

Translate this solution to a number line graph:

5. A linear inequality has been entered on the calculator so that the left side is expressed as Y1 and the right side as Y2.
a. What is the critical point as displayed in the TABLE of values?____

X	Y1	Y2
-2	0	12
-1	2	11
0	4	10
1	6	9
2	8	8
3	10	7
4	12	6

X=-2

b. What is the solution (expressed as a number line graph) to the inequality Y1 < Y2?

Directions: Solve each inequality graphically using the INTERSECT feature. Begin in a ZStandard WINDOW and adjust the viewing window accordingly. Critical points should be expressed in fraction form.

6. $3x - 8 < \frac{4}{5}(3x - 2)$

Xmin = ______ Xmax = ______

Ymin = ______ Ymax = ______

Solution Set:______________________________ Interval Notation:___________

Translate this solution to a number line graph:

7. $8+\frac{8}{5}x \geq \frac{1+9x}{8}$

Xmin = ______ Xmax = ______

Ymin = ______ Ymax = ______

Solution Set:______________________________ Interval Notation:__________

Translate this solution to a number line graph:

8. $-\sqrt{288}+x>2\left(2x-6\sqrt{2}\right)$

Xmin = ______ Xmax = ______

Ymin = ______ Ymax = ______

Solution Set:______________________________ Interval Notation:__________

Translate this solution to a number line graph:

✍9. Solve the combined inequality -4 < 4 + 2x < 8 graphically. Use the steps outlined in Exercise 1 as a model. Record any additional steps required to obtain a valid solution. Display your graph at the right.

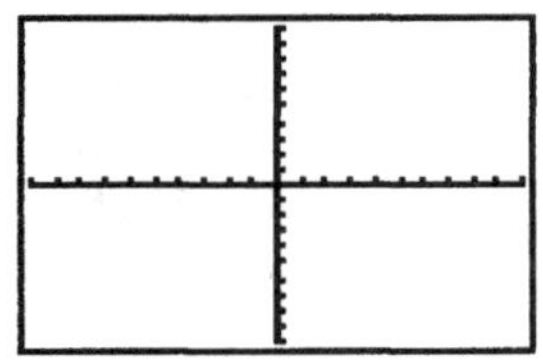

Solution Set:__________________________

Translate this solution to a number line graph:

10. The graph displayed at the right is the graph of the linear equation Y = P where P is a linear polynomial expression. The coordinates displayed represent the point where the line intersects the *x*-axis (*x*-intercept). Solve the indicated inequalities using the displayed graph and record your solution on the number line provided.

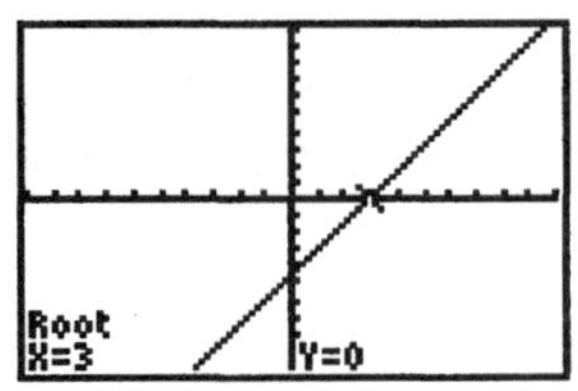

a. $y = 0$

b. $y > 0$

c. $y < 0$

d. $y \leq 0$

e. $y \geq 0$

THE FOLLOWING PROVIDES INFORMATION ABOUT THE GRAPH STYLE ICON WHICH YOU MAY WANT TO EXPERIMENT WITH WHEN GRAPHING MORE THAN ONE POLYNOMIAL AT A TIME.

TI-83/84series

The "graph style icon" is a feature that allows you to distinguish between the graphs of equations. Press **[Y=]** and observe the "\" in front of each Y. Use the left arrow to cursor over to this "\". The diagonal should now be moving up and down. Pressing **[ENTER]** once changes the diagonal from thin to thick. The graph of Y1 will now be displayed as a thick line. Repeatedly pressing **[ENTER]** displays the following:

◥: shades above Y1
◣: shades below Y1
-o: traces the leading edge of the graph followed by the graph
o: traces the path but does not plot
⋱: displays graph in dot, not connected MODE

The graphing icon takes precedence over the **MODE** screen. If the icon is set for a solid line and the **MODE** screen is set for DOT and not solid, the graphing icon will determine how the graph is displayed. Pressing **[CLEAR]** to delete an entry at the Y= prompt will automatically reset the graphing icon to default, a solid line.

TI-86

THE "GRAPH STYLE ICON" ALLOWS YOU TO DISTINGUISH BETWEEN THE GRAPHS OF EQUATIONS. PRESS **[GRAPH] [F1](Y(X)=)** AND OBSERVE THE "\" IN FRONT OF EACH Y. PRESS **[MORE]** FOLLOWED BY **[F3](STYLE)**. THE DIAGONAL SHOULD NOW HAVE CHANGED FROM THIN TO THICK. THE GRAPH OF Y1 WILL BE DISPLAYED AS A THICK LINE. REPEATEDLY PRESS **[STYLE]** TO DISPLAY THE FOLLOWING:

◥: SHADES ABOVE Y1
◣: SHADES BELOW Y1
-O: TRACES THE LEADING EDGE OF THE GRAPH FOLLOWED BY THE GRAPH
O: TRACES THE PATH BUT DOES NOT PLOT
⋱: DISPLAYS GRAPH IN DOT, NOT CONNECTED MODE

IT IS IMPORTANT TO REMEMBER THAT THE GRAPHING ICON TAKES PRECEDENCE OVER THE **MODE** SCREEN. IF THE ICON IS SET FOR A SOLID LINE AND THE **MODE** SCREEN IS SET FOR DOT AND NOT SOLID, THE GRAPHING ICON WILL DETERMINE HOW THE GRAPH IS DISPLAYED. PRESSING **[CLEAR]** TO DELETE AN ENTRY AT THE Y= PROMPT WILL AUTOMATICALLY RESET THE GRAPHING ICON TO DEFAULT, A SOLID LINE.

Solutions to Exercise Sets: **1.** {x|x > 1} 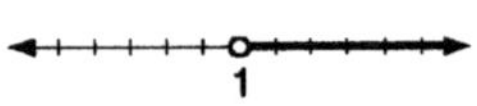

2. {x|x≥1.4} 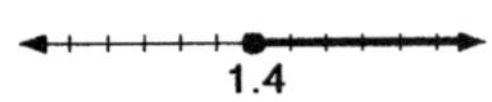

3. {x|x≥5} 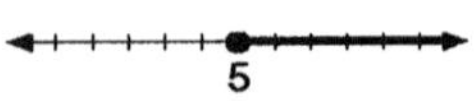

4. {x|x<2.25} 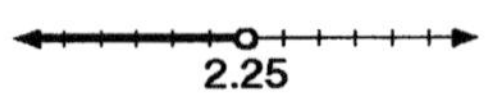

5. The critical point is 2 and the number line graph of the solution is 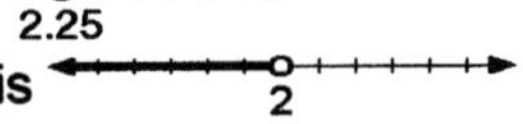

6. {x|x<32/3}, (-∞, 32/3) 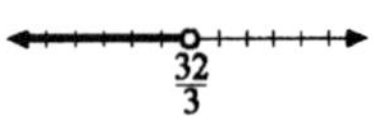

7. {x|x ≥ -315/19}, [-315/19, ∞) 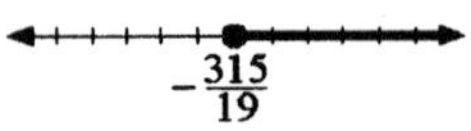

8. {x|x < 0}, (-∞, 0) 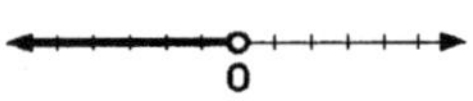

9. Solution Set: {x| -4 < x < 2}

10. **a.** {3} **b.** (3, ∞) **c.** (- ∞, 3) **d.** (-∞, 3] **e.** [3, ∞)

UNIT 11
GRAPHICAL SOLUTIONS: ABSOLUTE VALUE INEQUALITIES

*Prerequisites: Units #7 and #10

This unit continues to examine inequalities through the interpretation of graphs and tables.

Algebraically	**Numerically**	**Verbally**
$\|x + 3\| \le 6$ $-6 \le x + 3 \le 6$ $-9 \le x \le 3$ When x is replaced by any real number between negative nine and three the original inequality is a true statement.	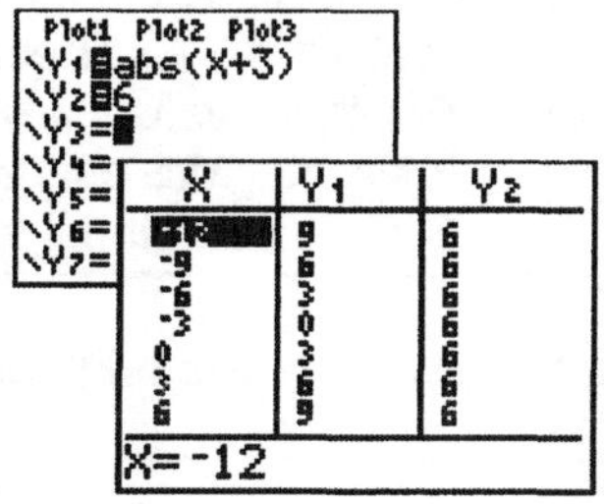	The table display confirms our algebraic solution. For the critical points, -9 and 3, we see that Y1=Y2. For the x-values smaller than -9 (-12 is displayed), Y1 > Y2, therefore -12 is not a solution. We assume all x-values smaller than -9 are not solutions. Between -9 and 3 we see that Y1 < Y2. These values **are** solutions. For the x-value larger than 3 (6 is displayed) we see that Y1 > Y2. We assume all x-values greater than 3 are not solutions.

Interpretation of the graphical display allows us to graphically solve the original inequality.

Graphically	**Interpretation**	**Notation**
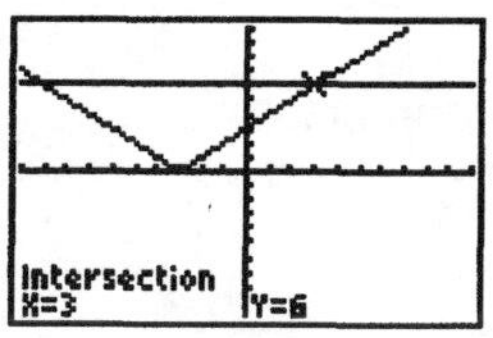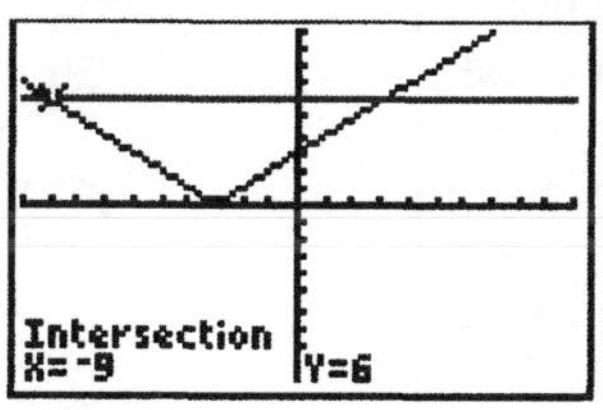 The INTERSECT feature of the graphing calculator was used twice to find the critical points. 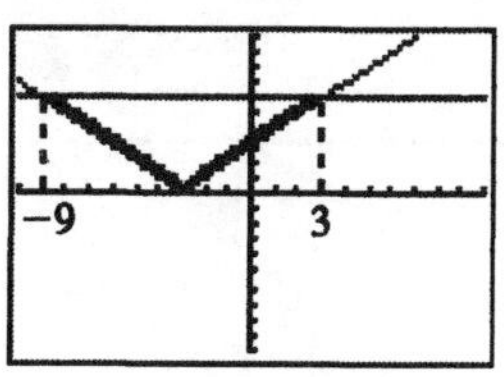	We know that Y1 < Y2 when the graph of Y1 is **<u>below</u>** the graph of Y2. We also know that the solutions to the inequality are **all** x values for which this is true. Press **[TRACE]** and be sure the TRACE cursor is on the graph of Y1 (the V-shape). As you TRACE along the graph of Y1 you will discover that the portion of Y1 that is less than Y2 is the portion of the graph of Y1 that is below the graph of Y2. This portion has been highlighted and should be between the two points labeled -9 and 3. The solutions to the **inequality** $\|x + 3\| < 6$ are all the x-values between -9 and 3 (the critical points). The solutions to the **equation** $\|x + 3\| = 6$ are the two critical points -9 and 3. Combining this information, we get the solution of $\|x + 3\| \le 6$ to be $-9 \le x \le 3$.	Solution Set: $\{x\| -9 \le x \le 3\}$, Number line graph:  Interval Notation: [-9,3]

The following illustrates the interpretation of graphs and tables to solve an absolute value inequality that is an *or* statement.

Algebraically	**Numerically**	**Verbally**
$\lvert x + 3\rvert > 6$ $x + 3 > 6$ or $x + 3 < -6$ $x > 3$ or $x < -9$ When x is replaced by any real number greater than 3 or less than -9, the original inequality is a true statement.	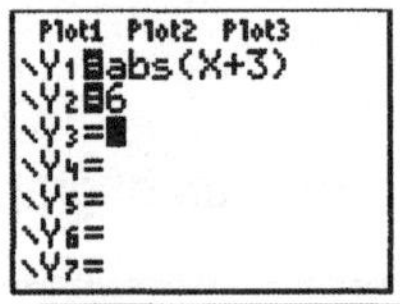 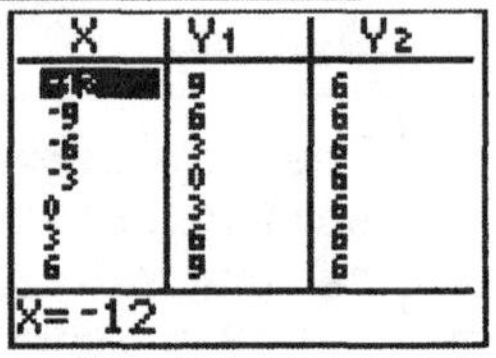	The table display confirms our algebraic solution. For the critical points, -9 and 3, we see that Y1=Y2. These numbers **are not** part of the solution. For the x-values smaller than -9 (-12 is displayed), Y1 > Y2, therefore -12 **is** a solution. We assume all x-values smaller than -9 are also solutions. Between -9 and 3 we see that Y1 < Y2. These values **are not** solutions. For the x-value larger than 3 (6) we see that Y1 > Y2. We assume all x-values greater than 3 **are** also solutions.

Interpretation of the graphical display allows us to graphically solve the original inequality.

Graphically	**Interpretation**	**Notation**
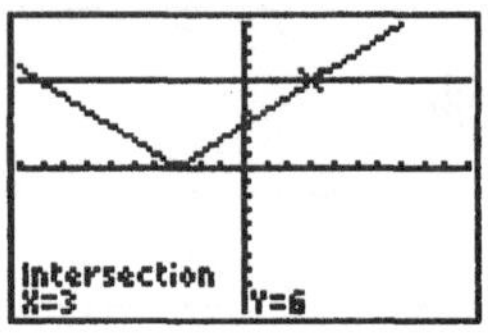 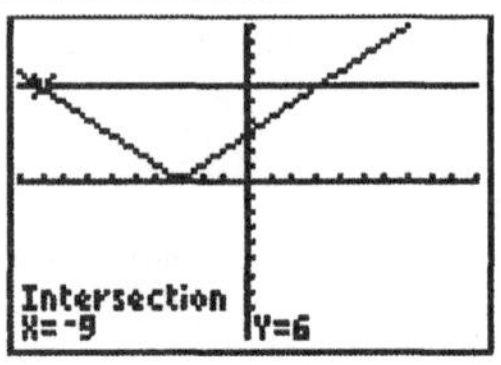 The INTERSECT feature of the graphing calculator was used twice to find the critical points. 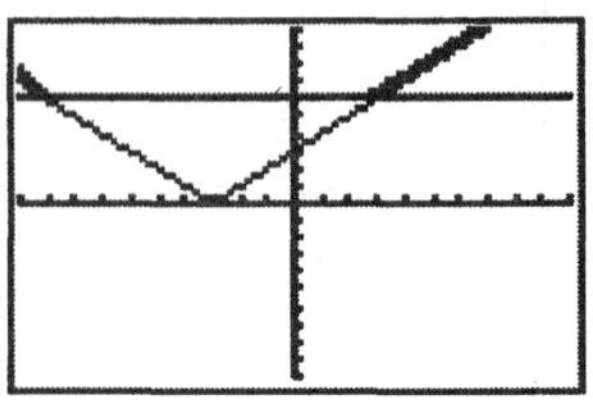	We know that Y1 > Y2 when the graph of Y1 is <u>**above**</u> the graph of Y2. We also know that the solutions to the inequality are **all** x values for which this is true. Press **[TRACE]** and be sure the TRACE cursor is on the graph of Y1 (the V-shape). As you TRACE along the graph of Y1 you will discover that the portion of Y1 that is greater than Y2 is the portion of the graph of Y1 that is above the graph of Y2. This portion has been highlighted and should be to the left and right respectively of the points $x = -9$ and $x = 3$. The solutions to the **inequality** $\lvert x + 3\rvert > 6$ are all the x-values to the left of -9 and to the right of 3 (the critical points). The solutions to the **equation** $\lvert x + 3\rvert = 6$ are the two critical points -9 and 3. Combining this information, we get the solution of $\lvert x + 3\rvert > 6$ to be $x < -9$ or $x > 3$.	Solution Set: $\{x \mid x < -9 \text{ or } x > 3\}$ Number line graph: 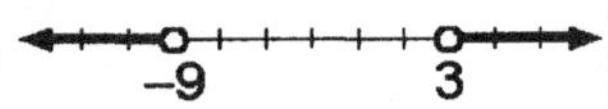 Interval Notation: $(-\infty,-9) \cup (3,\infty)$

Steps for Solving an Absolute Value Inequality Graphically

1. Enter the left side of the inequality at Y1 and the right side at Y2. You may want to use your graph style icon to differentiate between the two graphs.
2. Display the graph and determine the points of intersection of Y1 and Y2 (using your INTERSECT feature). These critical points represent the solution to the equation, Y1 = Y2.
3. Copy the intersect screen and draw vertical dotted lines from the critical points to the *x*-axis. Label these critical points on the *x*-axis (as illustrated in the graphical display on the first page of the unit).
4. Compare the graphs of Y1 and Y2 to determine where Y1 < Y2 or Y1 > Y2, depending on the problem and shade the appropriate portion of the graph of Y1.
5. Express your solution on a number line graph.
6. Check your solutions with the TABLE feature.

EXERCISE SET

Directions: Graphically solve each of the following inequalities following the steps that were outlined. Record the solution in set notation, as a number line graph and in interval notation.

1. $|2x - 1| \geq 5$

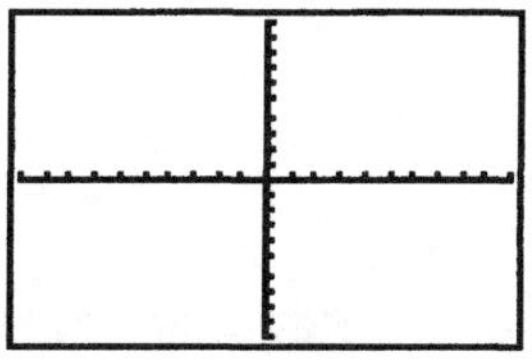

Solution Set:______________

Number line graph:

Interval Notation:__________

2. $\left|\frac{1}{2}x - 1\right| < 4$

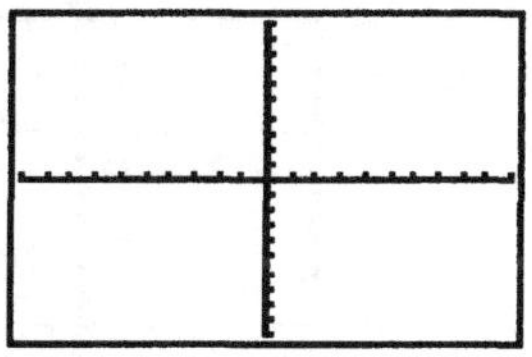

Solution Set:______________

Number line graph:

Interval Notation:__________

3. $\left|\frac{2x+5}{3}\right| < 4$

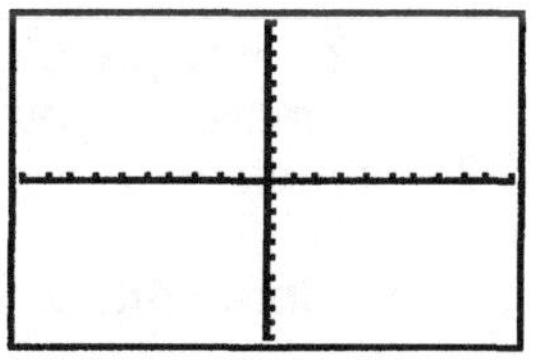

Solution Set:______________

Number line graph:

Interval Notation:__________

4. In your own words, explain what type of error could easily be made when graphing the expression $\left|\frac{2x+5}{3}\right|$ or $\left|\frac{1}{2}x - 1\right|$.

5. $|4x + 2| > -3$
This particular inequality represents a "special case." Carefully re-TRACE your highlighted portion of the graph before deciding on the solution.

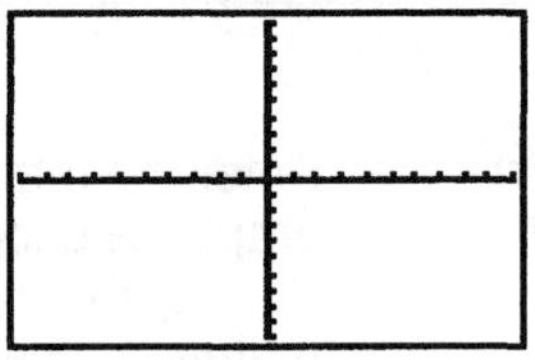

Solution Set:______________

Number line graph:

Interval Notation:___________

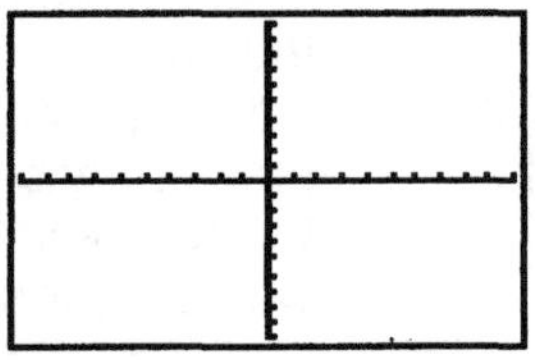

6. $|4x + 2| < -3$ (another "special case")

Solution Set:________

✍7. Consider why #5 and #6 are labeled as "special cases." Could #5 and #6 have been solved by merely "looking" at the inequality? Think carefully about the definition of absolute value before formulating your response.

✍8. Summarizing Results: Because the left and the right sides of equations and inequalities are graphed as separate expressions, the graphical representations of the solutions to each of the following problems all look alike.

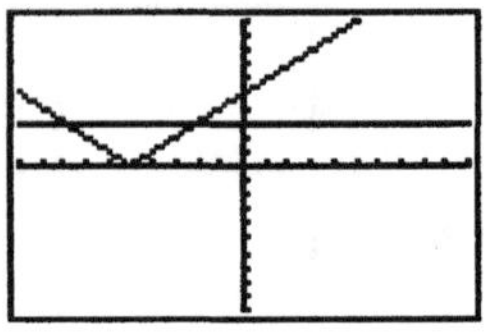

$|x + 5| = 3$

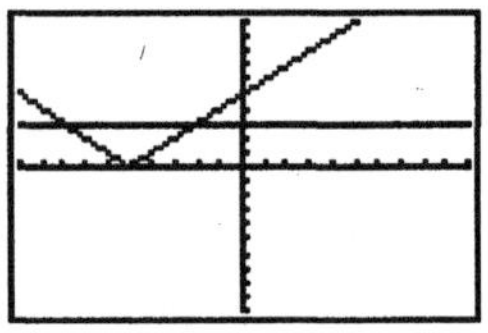

$|x + 5| < 3$

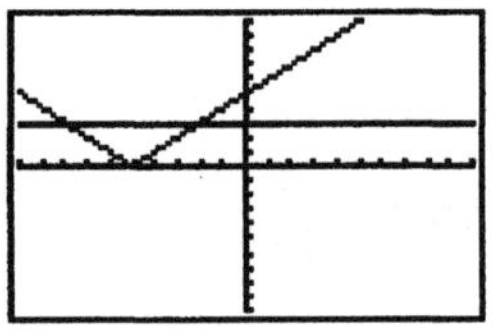

$|x + 5| > 3$

The interpretations of the solutions represented by the graphs above are all different. To summarize your results from this unit, explain how to interpret the solution represented by each graph.

a. Interpretation of $|x + 5| = 3$:

b. Interpretation of $|x + 5| < 3$:

c. Interpretation of $|x + 5| > 3$:

Solutions: **1.** $\{x \mid x \leq -2 \text{ or } x \geq 3\}$,

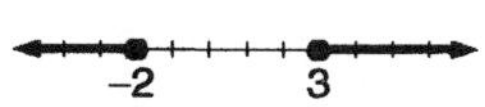

$(-\infty,-2] \cup [3,\infty)$

2. $\{x \mid -6 < x < 10\}$,

$(-6,10)$,

3. $\{x \mid -8.5 < x < 3.5\}$,

$(-8.5,3.5)$

4. You might not put parentheses around the numerator of the fractions or fail to enclose the entire fraction in parentheses when using the absolute value command.

5. $\mathbb{R}$ or $\{x \mid x \text{ is a real number}\}$, $(-\infty,\infty)$

6. null set

7. Remember that an absolute value is at least 0 or larger. Thus, an absolute value is always greater than a negative number for any value of the variable and an absolute is never less than any negative number regardless of the value of the variable.

8. a. The critical points are the solution. **b.** The graph of Y1 must be below the graph of Y2; therefore the solution is the *x*-values between the critical points. **c.** The graph of Y1 must be above the graph of Y2; therefore the solutions are all *x*-values that are less than the smaller critical point or greater than the larger critical point.

UNIT 12
GRAPHICAL SOLUTIONS: QUADRATIC INEQUALITIES

*Prerequisite: Units #8 and 10

This unit will graphically examine quadratic inequalities by using the ROOT/ZERO option of the calculator and by interpreting the relationship between the graphical displays of each side of the inequality. REMEMBER: To use the ROOT/ZERO there must be a zero on one side of the inequality.

Special Cases

The first five examples represent "special cases." They will be the quickest to solve of all the inequalities. All graphs will be displayed in the standard viewing window unless otherwise noted. Set this viewing window now.

Example	**Graphically**	**Interpretation**
1. Graphically solve $2x^2 - x + 1 > 0$.	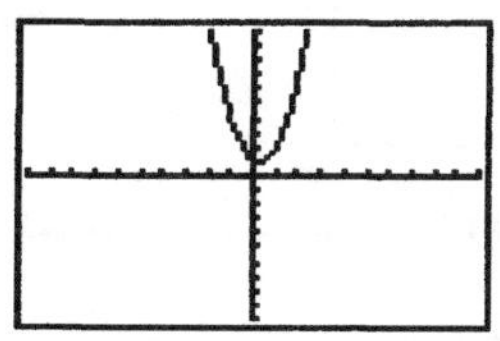 Enter the left side of the inequality at Y1 and the right side at Y2 . We want to know where Y1 > Y2. Since Y2 = 0 is the X-axis, we do not see it as a separate line. It is, however, graphed. *Note: From this point on, make a mental note that Y1 is always being compared to the x-axis and do not enter 0 at Y2.*	**Think** about where Y1 is greater than Y2 (i.e. the x-axis). It is greater than the x-axis where it is **above** the axis. Highlight the portion(s) of Y1 that is(are) above the x-axis. Since the graph of Y1 is entirely above the x-axis, any value of x is a solution to the original inequality. The solution set is $\mathbb{R}$ (the set of all real numbers).
2. Graphically solve $2x^2 - x + 1 < 0$	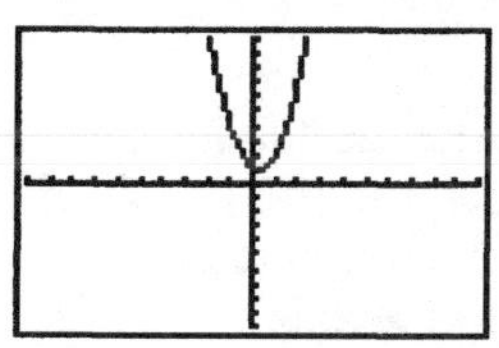 Enter the left side of the inequality at Y1 and the right side at Y2 . We want to know where Y1 < Y2, that is, where Y1 is below the x-axis.	Since the graph does not dip below the x-axis, there are no x-values for which the inequality is true. The solution is the empty set, ϕ or { }.
3. Graphically solve $x^2 + 4x + 4 \leq 0$	Enter $x^2 + 4x + 4$ after Y1. **TRACE** along the path of the curve. 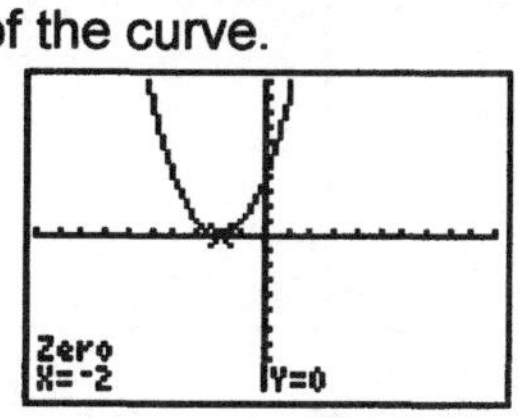 Use the ROOT/ZERO option to determine where the graph is ***equal to*** zero. There is only ***one*** valid solution: $x = -2$.	At what point(s) is the graph of $x^2 + 4x + 4$ ***less than*** 0? Tracing reveals that there are no values for x for which y is negative. The solution of the inequality $x^2 + 4x + 4 \leq 0$ is $x = -2$.

Example	Graphically	Interpretation
4. Graphically solve $x^2 + 4x + 4 \geq 0$	Enter $x^2 + 4x + 4$ after Y1. **TRACE** along the path of the curve. 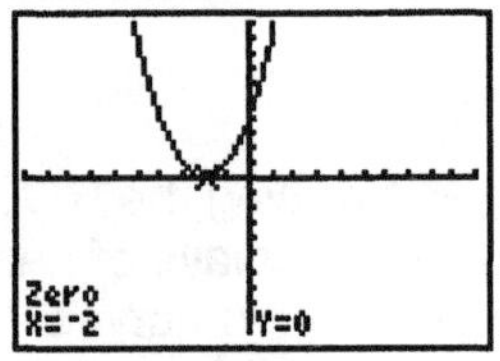 Use the ROOT/ZERO option to determine where the graph is ***equal to*** zero. $x = -2$.	At what point(s) is the graph of $x^2 + 4x + 4$ ***greater than*** 0? Tracing reveals that all values for x yield positive y-values . Therefore, the solution is $\mathbb{R}$.
5. Graphically solve $x^2 + 4x + 4 > 0$	Enter $x^2 + 4x + 4$ after Y1. **TRACE** along the path of the curve. 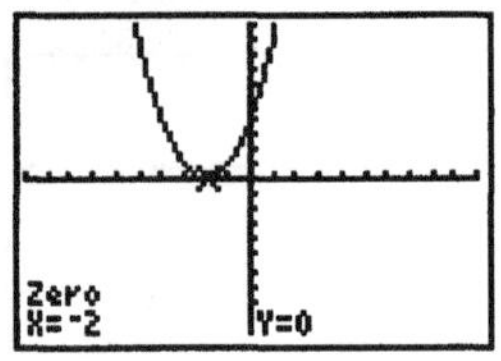	The graph of Y1 is greater than 0 everywhere ***except*** at the point where Y1 equals 0 ($x = -2$). Thus the solution is all real numbers ***except*** - 2. Using set notation: $\{x \mid x \neq -2\}$.

General Inequalities

In Example 3, $x^2 + 4x + 4 \leq 0$, the critical point in the solution set was $x = -2$. At $x = -2$, $x^2 + 4x + 4 = 0$. Negative two is a root of the equation. When solving inequalities, the critical points (the roots of the corresponding equation) will be the endpoints of the interval(s) of the solution region(s). The following example illustrates the procedure for solving quadratic inequalities. The way the graphical display is interpreted determines the solution to a given equation or inequality. The algebraic statement indicates that the trinomial is less than **OR** equal to zero. Remember, 0 is represented by the x-axis. We will illustrate the steps for solving quadratic inequalities in the chart below, by solving $x^2 - x - 6 < 0$.

Graphically	Interpretation	Notation
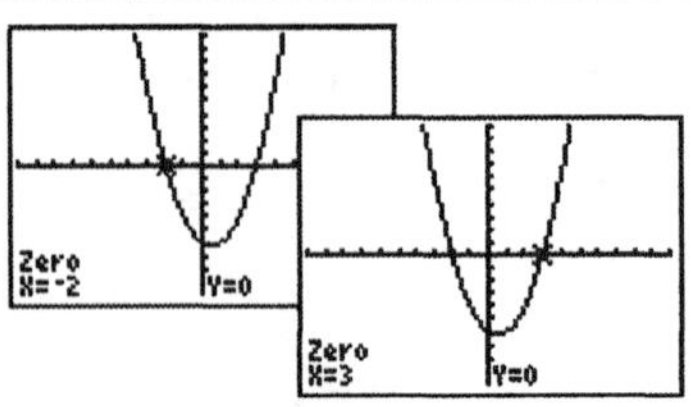 The ROOT/ZERO feature was used twice to find the critical points. 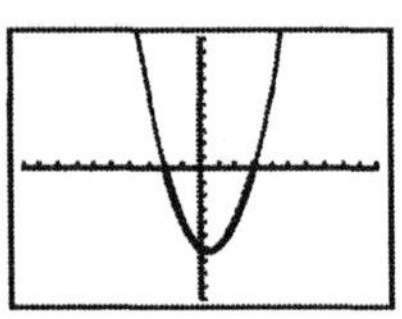	Because the original inequality was less than, we know that the graph of Y1 is less than Y2 when it is *below* the graph of Y2 (the x-axis). The critical points, -2 and 3, are *not* a part of the solution because the inequality did *not* include equality. The graph shows that Y1 < Y2 for all x-values between $x = -2$ and $x = 3$. The TABLE confirms our solution. 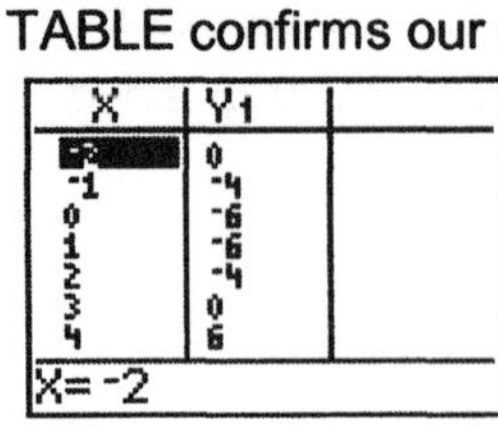	The solution set is the set of all x values that yield the highlighted section of the graph. That is, $\{x \mid -2 < x < 3\}$. The corresponding number line graph is  In interval notation, the solution would be specified as (-2, 3).

To solve $x^2 - x - 6 > 0$, we interpret Y1 > Y2 as the portion of the graph above the x-axis (in two pieces). The x-values that are the solutions are those to the left of the critical point -2 and to the right of the critical point 3. This is expressed in set notation as $\{x \mid x \leq -2 \text{ or } x \geq 3\}$; as 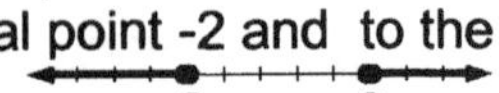 using the number line, and $(-\infty, -2) \cup (3, \infty)$ in interval notation.

Steps for Solving Quadratic Inequalities	
1.	Be sure that the right hand side of the inequality is zero.
2.	Enter the polynomial at the Y1 prompt.
3.	Display the graph in the standard viewing window. Copy this display on your paper.
4.	Circle the *x*-intercepts (zeroes or roots) of the graph. These are the critical points. The circles will remain *open* if the inequality is strictly > or < and will be *closed* if equality is included (the inequality is $\geq$ or $\leq$).
5.	Use the ROOT/ZERO option (under the **CALC** menu) to determine the values of the *x*-intercepts (TI-86 users press **[GRAPH]** **[MORE]** **[F1](MATH)** to locate the ROOT option). Label the values of these two critical points on the display.
6.	Highlight the appropriate section of your graph (the part above the *x*-axis for the inequality > and the part below the *x*-axis for the inequality <).
7.	Record the solution using set, interval, or graph notation.

Note: The TABLE feature of the calculator may be helpful in determining solution regions once the critical points have been calculated.

EXERCISE SET

Use the steps outlined to graphically solve each of the following inequalities. For each problem sketch the graphical display and label the critical points. Record your solution in the following forms:
a. solution set b. number line graph c. interval notation

1. $2x^2 - x - 10 \leq 0$

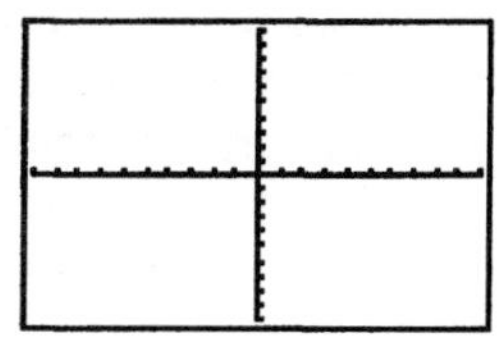

a. Solution Set:______________________________

b. Number line graph:

c. Interval Notation:______________________

2. $3x^2 + x - 4 \geq 0$ (Record critical points as fractions.)

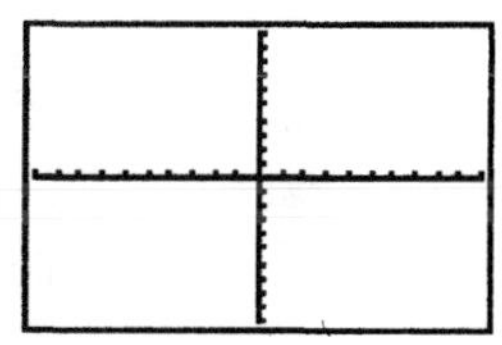

a. Solution Set:______________________________

b. Number line graph:

c. Interval Notation:______________________

3. $3x^2 + x - 4 < 0$ (Record critical points as fractions)

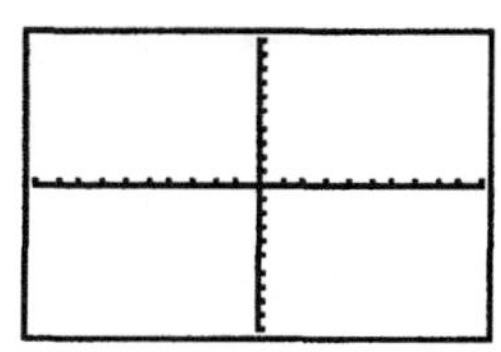

a. Solution Set:______________________________

b. Number line graph:

c. Interval Notation:______________________

4. $x^2 + 10x + 25 \geq 0$

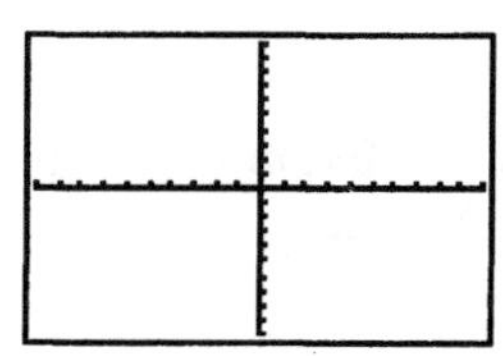

a. Solution Set:______________________________

b. Number line graph:

c. Interval Notation:______________________

5. $x^2 + 10x + 25 > 0$

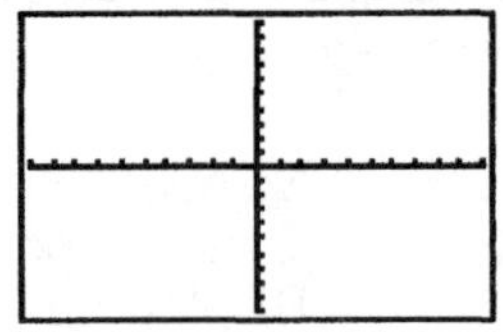

a. Solution Set:________________________

b. Number line graph:

c. Interval Notation:____________________

6. $4x^2 - 12x < -9$

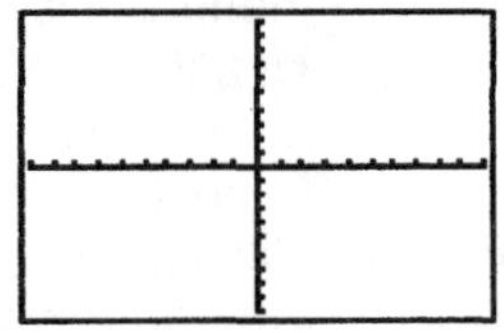

Solution Set:________________________

7. $4x^2 - 12x + 9 > 0$

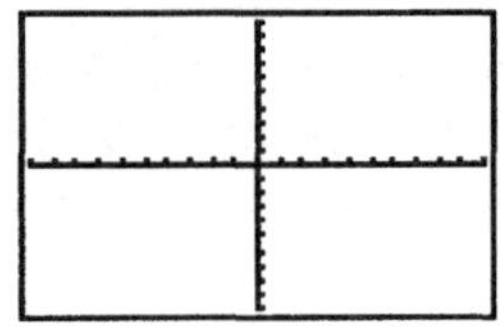

a. Solution Set:________________________

b. Number line graph:

c. Interval Notation:____________________

NOTE: When using ROOT/ZERO the *x*-value may be a decimal approximation because it is dependent upon the placement of the bound markers. Therefore, two students could get two different approximations for *x*. If one student gets an approximation of 1.4999998 should they conclude this is the critical point or is it 1.5? Attempting to convert to a fraction would seem to indicate that it is not 1.5. However, since the critical point(s) is the solution to the equation $4x^2 - 12x + 9 = 0$, then the critical point should be the *x*-value in the TABLE where Y1 = Y2. This can be accomplished in three ways:

a. Set the TABLE to start at $x = 1.5$ (the table increment is not critical since only one value is being checked). We want to know if Y1 = Y2 when $x = 1.5$. Press **[2nd] <TblSet>**, start the table at 1.5, and press **[2nd] <TABLE>**. At $x = 1.5$, Y1 = Y2. Thus $x = 1.5$ is the exact critical point; $x = 1.4999998$ is an approximation.

b. Use VALUE (EVAL) to determine if Y1 = 0 when $x = 1.5$

When using the graph screen to solve equations/inequalities, you should be aware that the coordinate values displayed approximate the actual mathematical coordinates. The accuracy of these display values is determined by the height and width of the pixel space being displayed. The space height/width formulas are discussed in detail in the unit entitled "Calculator Viewing Windows."

✍8. Explain the similarities and differences between graphically solving equations and inequalities.

✍9 Explain the significance of the critical points.

Solutions: **1.** $\{x|-2 < x \leq 2.5\}$,

$[-2, 2.5]$

2. $\{x|\ x \leq -4/3 \text{ or } x \geq 1\}$, 

$(-\infty,-4/3] \cup [1, \infty)$

3. $\{x|-4/3 < x < 1\}$, 

$(-4/3, 1)$

4. $\mathbb{R}$, $(-\infty, \infty)$ **5.** $\{x|x \neq -5\}$, 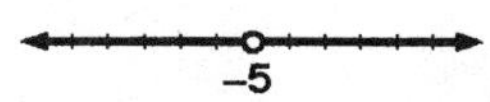

$(-\infty, -5) \cup (-5, \infty)$

6. Null Set **7.** $\{x|\ x \neq 1.5\}$, 

$(-\infty, 1.5) \cup (1.5, \infty)$

8. Answers may vary. **9.** Answers may vary.

UNIT 13
GRAPHICAL SOLUTIONS: RATIONAL INEQUALITIES

* Prerequisite: Unit #12

The previous unit examined solving quadratic inequalities. To solve rational nonlinear inequalities we will use some of the same procedures from the previous unit and investigate necessary modifications.

Example 1	Graphically
Graphically solve $\frac{12}{x} \geq 3$.	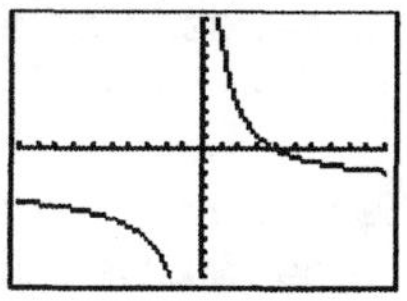 Rewrite the inequality as $\frac{12}{x} - 3 \geq 0$. (Comparing an expression to zero allows the use of the *x*-axis as a reference point.) Enter $\frac{12}{x} - 3$ at the Y1 prompt and view the graph in the standard viewing window. 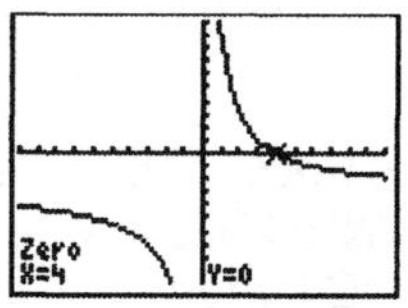 Find the critical point(s). Circle the point where the graph crosses the *x*-axis. Use the ROOT/ZERO option to find the critical point. Shade the part of the graph that corresponds to Y1 > 0. 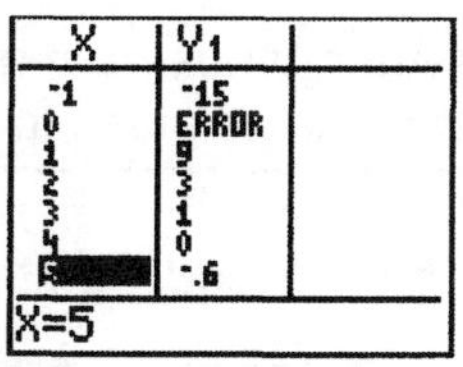 To confirm that 4 is a root (critical point), access the TABLE (first set the table to begin at - 1 and increment by 1). When *x* = 4, Y1 has a value of 0. This confirms that 4 is the root (and thus a critical point). However, notice that when *x* = 0 that Y1= ERROR. There is an ERROR message for this value of *x*. Because the fraction $\frac{12}{x}$ is undefined when *x* = 0, this value for *x* *is not* part of the solution. This is an excluded (or restricted) value.

Solution:

a. Locate the excluded value (0) and the root (4) on the graph.

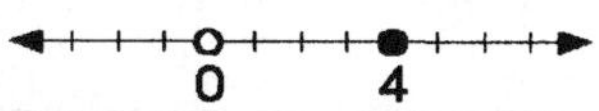

Notice that the coordinate 0 is marked with an open circle, while the coordinate 4 is marked with a closed circle. These are different because 4 is a ***solution*** to the inequality, whereas the value 0 ***cannot*** be a solution to the inequality because it results in an expression that is undefined.

b. **Shade** the part(s) of the pictured number line that correspond to those points of the graph that are above the *x*-axis.

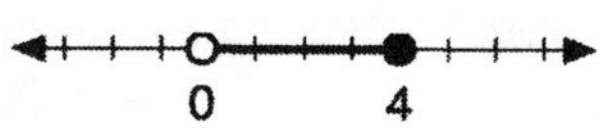

Your graph should correspond to $\{x \mid 0 < x \leq 4\}$ which is expressed as (0,4] in interval notation.

c. Accessing the TABLE feature allows us to check the solution. Notice in the displayed TABLE above, that for values of *x* *smaller* than 0, the Y1 values are negative, and are not a part of the desired solution. When *x* is *greater* than 0 but *less* than 4 the corresponding Y1 values are positive, and thus are solutions to $\frac{12}{x} - 3 \geq 0$. Scrolling past 4 (to *x* values *greater* than 4) the corresponding Y1 values are negative, and are not a part of the desired solution.

Example 2	**Graphically**
Graphically solve the inequality $\frac{-8}{x-4} \le \frac{5}{4-x}$.	Rewrite the inequality so that one side equals zero: $\frac{-8}{x-4} - \frac{5}{4-x} \le 0$ 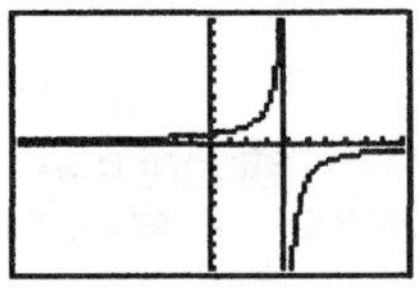 Enter the left side at the Y1 prompt and press **[GRAPH]**. Locate the excluded value of 4 by setting denominator factors equal to 0 and solving for *x*. Find any critical point(s). These are the point(s) where the graph crosses the *x*-axis. Use the ROOT/ZERO option to **try** to find all the root(s). 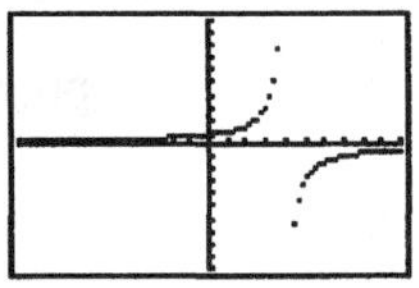 To better see what is going on, put the calculator in **DOT** mode. To do this, press **[MODE]** and cursor down to **Connected** and then right to **Dot**. Press **[ENTER]** to highlight this mode. (You could also use your graph style icon - see Unit #10) Press **[GRAPH]**. When comparing the two graphical displays notice the vertical line (where we believed there was a root) is gone. In CONNECTED mode this line connected two adjacent pixel points on the graph. In DOT mode it is clear that these two points are should not be connected. To connect them would mean that 4 is an acceptable value for *x*. The graph never crosses the *x*-axis but rather jumps from a location above the *x*-axis to one below it. **TRACE** and observe the *x*-values to confirm this.

Solution:. We want the solutions to the inequality $\frac{-8}{x-4} - \frac{5}{4-x} \le 0$. Place 4 on the number line and circle it (an excluded value) and shade to the right:

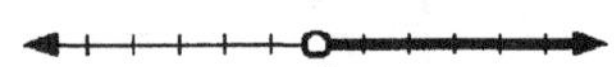

Use your TABLE to confirm that *x*-values *greater* than 4 are solutions. The solution set is $\{x \mid x > 4\}$ or $(4,\infty)$ in interval notation.

The following steps outline the method of finding the solution(s) to a rational inequality.

Solving Rational Inequalities Graphically

1. Rewrite the inequality with one side equal to 0. Enter the non-zero side at the Y1 prompt. Remember, the calculator should be in DOT MODE or the dot style icon should be activated.

2. Find the excluded values by setting each denominator equal to 0 and solving for *x*. Enter these values on the number line and mark them with open circles so that they are not inadvertently included in the solution.

3. Find the roots/zeroes of the equation associated with the inequality by using the ROOT/ZERO option. Enter the roots on the number line and mark them with closed circles if the inequality is $\le$ or $\ge$, and with open circles if the inequality is $<$ or $>$.

4. Shade the regions on the number line that represent the *x*-values of the ordered pairs on the graph where the y-values are greater than/less than 0 as determined by the inequality. You may also determine the appropriate *x*-values by testing values in each region of the number line. You can accomplish the same thing by using the TABLE.

5. Express the solution as a set and in interval notation.

EXERCISE SET

Directions: Use the steps outlined above to solve each inequality below. Begin by setting the calculator in Connected MODE with a standard viewing WINDOW. Remember to access the TABLE feature (when appropriate) and convert to dot mode as **YOU** deem appropriate. Use the combination of algebra and calculator that makes you feel comfortable.

1. $\dfrac{x^2+6x+9}{x+5} > 0$

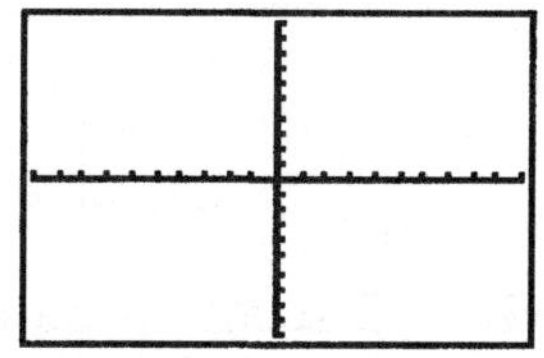

   a. Number line graph:

   b. Solution set: ____________________

   c. Interval notation: ____________________

2. $\dfrac{2}{x-2} < \dfrac{3}{x}$

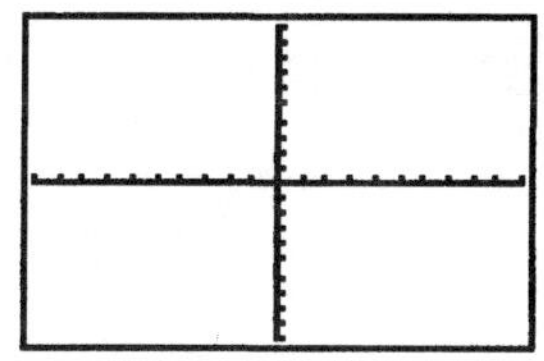

   a. Number line graph:

   b. Solution set: ____________________

   c. Interval notation: ____________________

3. $\dfrac{(2x-1)(x-5)}{x+3} \le 0$

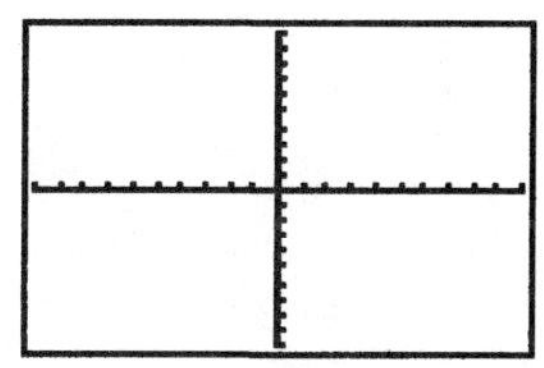

   a. Number line graph:

   b. Solution set: ____________________

   c. Interval notation: ____________________

Note: In #3, did you shade <u>only</u> between the roots of 1/2 and 5? This problem illustrates an important point: *always* look to the left (or right) of the restricted values and/or critical points to see how the graph behaves.

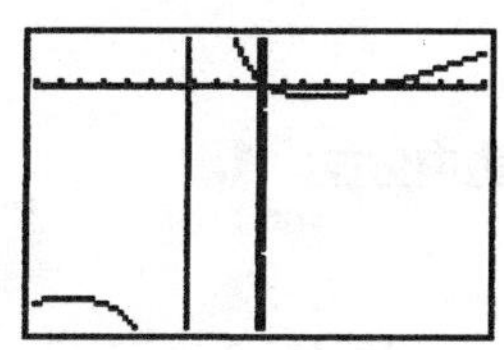

Set the calculator in Connected MODE.
TRACE left on the graph; go beyond the vertical line that connects the two non-adjacent pixel points. You will not see any more graph (TRACE until your graph shifts <u>at least</u> once).
Look at the *x*- and *y*-coordinates at the bottom of the screen, and then check the window values.
Press **[WINDOW]** and scroll down to Ymin and enter -50.
Press **[GRAPH]** and compare your display to the one pictured.

The display should help you understand why the region of the number line less than -3 is shaded.

4. Explain what excluded values are, how to find them, and their inclusion or non-inclusion in solutions.

5. Discuss the advantages and disadvantages of looking at solutions of non-linear rational inequalities in both Dot and Connected MODE.

Solutions: **1.** 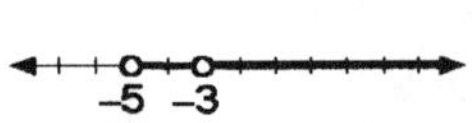

$\{x|-5 < x < -3 \text{ or } x > -3\}$ $(-5,-3) \cup (-3,\infty)$

2. 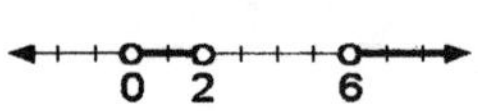

$\{x|\ 0 < x < 2 \text{ or } x > 6\}$ $(0, 2) \cup (6,\infty)$

3. 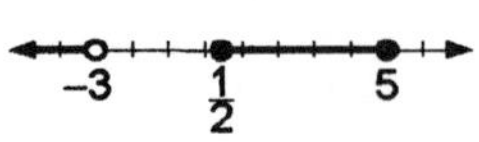

$\{x|x < -3 \text{ or } \frac{1}{2} \le x \le 5\}$ $(-\infty,-3) \cup [\frac{1}{2},5]$

4. Answers may vary. **5.** Answers may vary.

UNIT 14
GRAPHING BASICS

*Prerequisite: Unit # 3

The graphing calculator can be used to evaluate algebraic expressions.

Algebraically: Evaluate the polynomial $\frac{3}{4}x+6$ for $x = -8$, $x = -4$, $x = 0$, $x = 4$, and $x = 8$.

Substitute each of the given x-values into the polynomial and then simplify the polynomial by following the order of operations.

when x = - 8: $\frac{3}{4}(-8)+6=0$; when x = - 4 : $\frac{3}{4}(-4)+6=3$; when x = 0: $\frac{3}{4}(0)+6=6$;

when x = 4: $\frac{3}{4}(4)+6=9$; when x = 8: $\frac{3}{4}(8)+6=12$

Graphing Calculator: The STOre feature could be used repeatedly to evaluate the expression for the different values of x from the home screen.

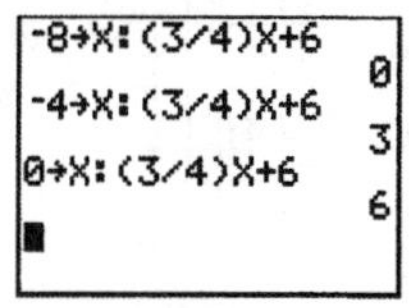

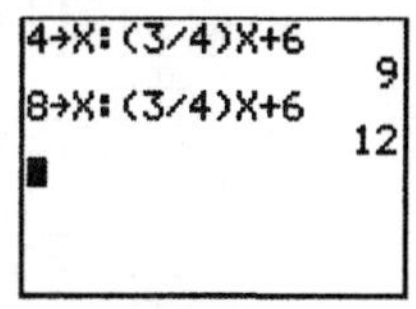

The TABLE feature can also be used to evaluate the polynomial for the given values of x.

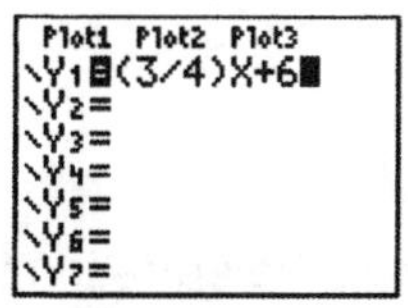

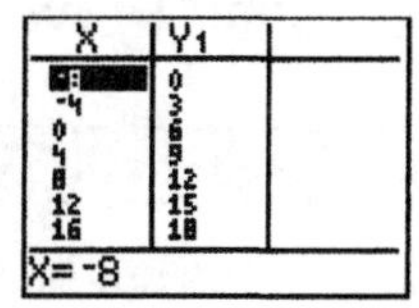

The TABLE was set to begin at the smallest desired value of x (- 8) and was incremented by 4 so that all the desired values would be displayed on a single TABLE screen.
Because the polynomial was stored at the Y1 slot, the evaluation of the polynomial for the given values of x are displayed in the Y1 column of the table. (Refer to Unit #3 if you need keystroking information.)

Notice, when we store the *expression* on the *y*-edit screen (Y=) we have an *equation* in two variables, x and y. After inputting a value for x we get an output value for y. This can be written as an ordered pair, (x , y).

Algebraically

Plot the ordered pairs (- 8, 0), (-4, 3), (0, 6), and (4, 9).
Connect the points graphed by drawing a straight line through those points (use a ruler).
Put arrowheads at the each end to indicate that the line goes on indefinitely.

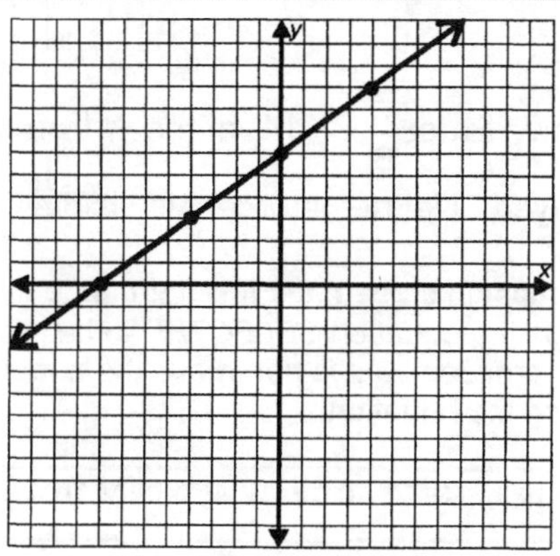

Hand Drawn Graph

Graphing Calculator

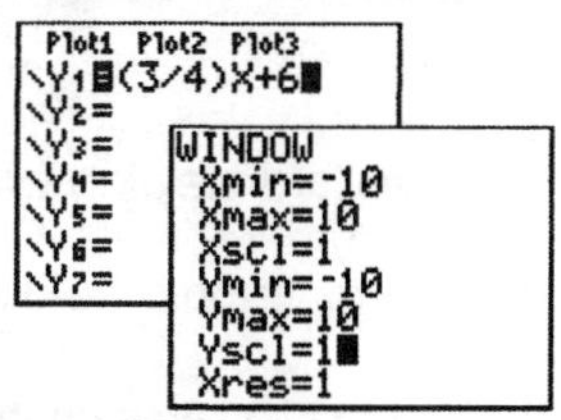

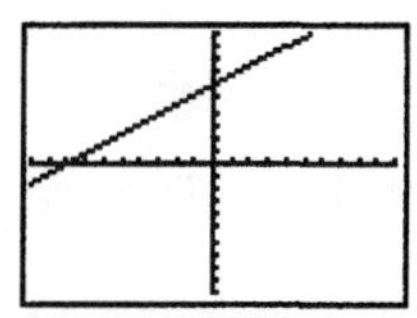

Standard Window

Keystrokes:
Press **[WINDOW]** (TI-86 users press **[Graph] [F2] (WIND)**) and enter -10 for Xmin, 10 for Xmax, -10 for Ymin, and 10 for Ymax. The Xscl and Yscl should both equal 1. The resolution (Xres) should also equal 1 and never be changed. This is the standard viewing window.

The standard viewing window can also be set quickly by pressing **[ZOOM] [6] (6:ZStandard)** (TI-86 users press **[GRAPH] [F3] (ZOOM) [F4](ZSTD)**). Any time you want to see your window values (x and y maximum and minimum values and scale) press **[WINDOW]** (TI-86 users press **[GRAPH] [F2](WIND)**).

The calculator drawn graph is not a *straight* line as expected. This is because the graph screen is made up of small boxes called pixels. When any part of a graph is in a pixel, the entire pixel lights up. If we shaded the pictured hand drawn graph (shading each square that contains any part of a line) we can get a good idea of how the calculator is programmed to graph.

The calculator plots points by lighting up little squares on the screen called pixels.

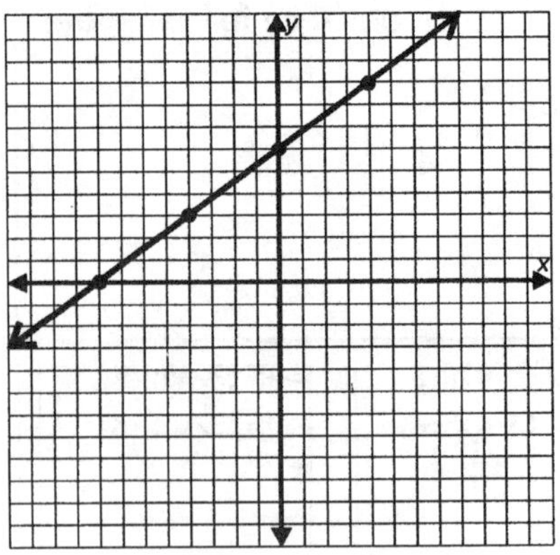

Hand Drawn Graph

The pictured graphs show the comparison between your hand drawn graph using the ordered pairs, the actual calculator display and a sketch where each square containing a piece of graph is shaded.

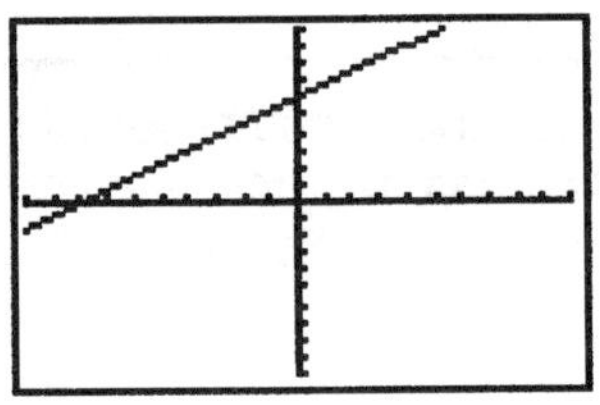

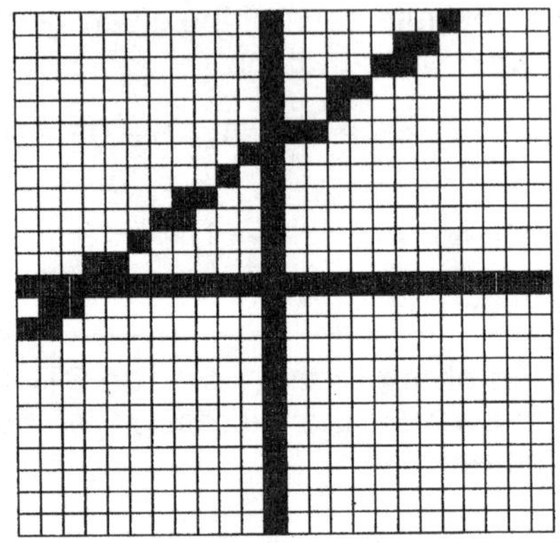
Pixel Sketch

Where Did the Graph Go?

Many times students are frustrated when the equation they have carefully keystroked into the **Y=** screen does not appear when **GRAPH** is pressed. What actually happens to the graph? Suppose you graphed $y = 2x^2 + 4x + 12$ on graph paper and then graphed this same equation on the calculator with the viewing window set to ZStandard. The figure at the right illustrates the handsketched graph with the section displayed on the ZStandard screen outlined in a bold black line. The viewing window selected is not large enough to display the graph. We will now look at options available to you that will give you the practice necessary to feel confident about setting the viewing window correctly to display a complete graph. A complete graph is defined to be one in which all x- and y-intercepts are visible as well as any peaks/maximums and valleys/minimums.

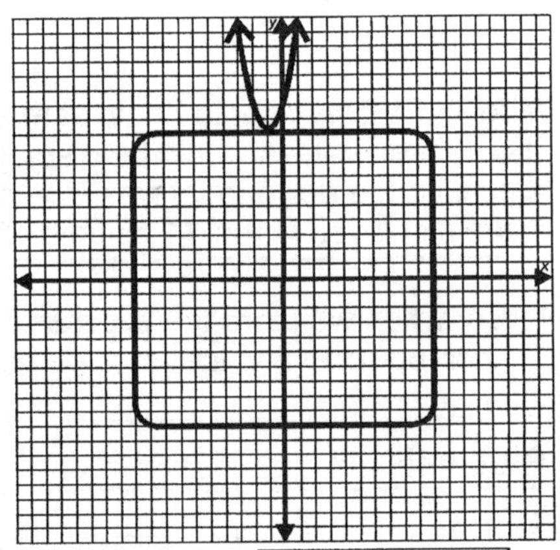

Before proceeding, set the standard viewing window. Enter $\frac{7}{6}x - 14$ at the Y1 prompt, **[GRAPH]**. The graph at the right should be displayed. Neither intercept is visible.

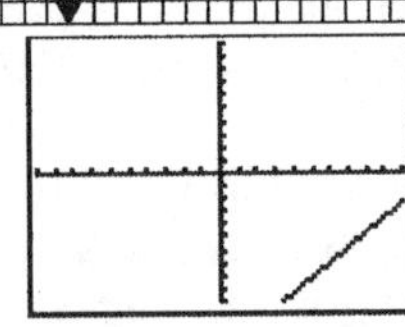

Note: If the equation entered is not displayed on the graphing screen, the first item to be checked is the entry of the equation on the **Y=** screen. Is the equation *SELECTED* to be graphed? That is, is the equal symbol beside the Y1 prompt highlighted? If it is, proceed. If it is not, move the cursor over the equal sign and press **[ENTER]** to highlight the equal sign, thus activating the equation.

TI-86	**[F5](SELCT)** ON THE **Y(x)=** MENU WILL BE USED TO ACTIVATE AND DEACTIVATE EQUATIONS.

If the equation is activated, then begin the process of adjusting the WINDOW by locating the x- and y-intercepts of the graph.

Finding the *y*-intercept

Algebraically

Substitute 0 in the equation for the variable, *x*:

$$y = \frac{7}{6}x - 14$$

$$y = \frac{7}{6}(0) - 14$$

$$y = -14$$

The *y*-intercept is - 14.

As an ordered pair, this would be written (0, -14).

Graphically

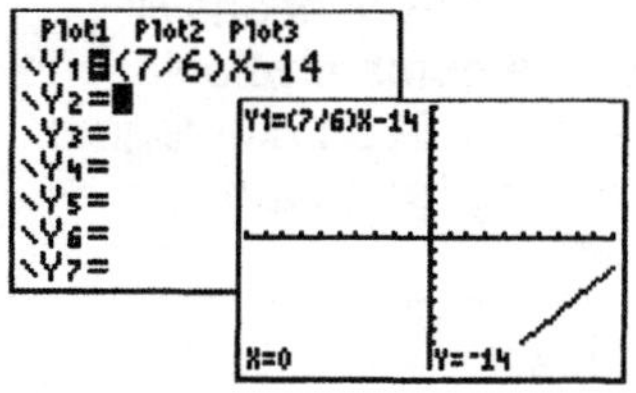

After pressing **[GRAPH]** to display the graph, press **[TRACE]** (TI-86 users press **[F4] (TRACE)**). The coordinates of the *y*-intercept, (0, - 14), are displayed at the bottom of the screen.

Caution: This must be done immediately after displaying the graph.

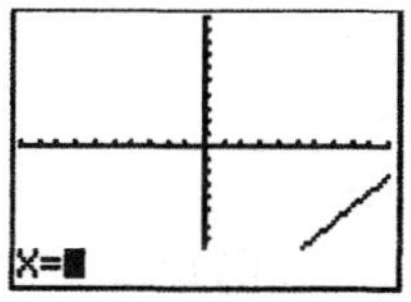

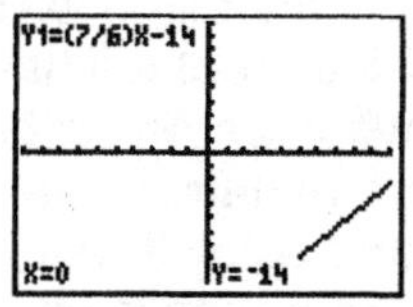

You may also use the VALUE feature on the calculator. With the graph displayed, press **[2nd] <CALC> [1] (1:value)** (TI-86 users - with line of menu displayed on the graph screen - press **[MORE] [MORE] [F1] (EVAL)**).
At the prompt on the graph screen, enter 0 (the *x*-value of the *y*-intercept). Press **[ENTER]** to see the intercepts displayed at the bottom of the screen.

Finding the *x*-intercept

Algebraically

Substitute 0 in the equation for the variable, *y*:

$$y = \frac{7}{6}x - 14$$

$$0 = \frac{7}{6}x - 14$$

$$0 = 7x - 84$$

$$84 = 7x$$

$12 = x$ The *x*-intercept is 12.

As an ordered pair, this would be written (12, 0).

Graphically

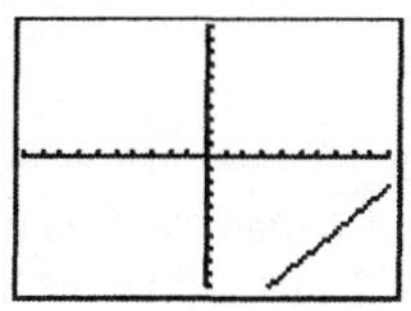

From the display (in the standard viewing window), we see that the *x*-intercept is to the right. We will extend our *x*-axis to the right by changing the Xmax on the WINDOW screen to 20 (a guess).

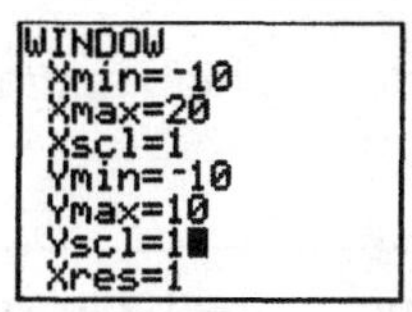

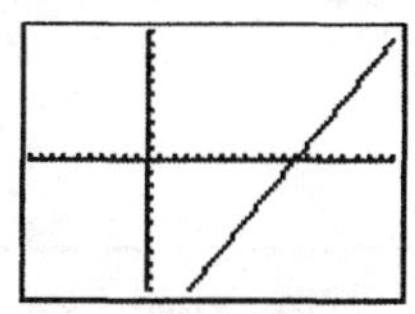

Note: To set a good window we need to be able to see the *x*-intercept. It is not necessary to graphically determine it.

To see both intercepts, we return to the WINDOW screen and set appropriate values so that the *y*-intercept of -14.On the WINDOW screen, use the down arrow key to move the cursor to Ymin and replace the Ymin with -20, a value smaller than -14. This smaller value allows a clear view of the *y*--intercept.

Pressing **[GRAPH]** displays the screen at the right. This a satisfactory graph because both the *x* and *y*-intercepts are displayed. Because the equation is a linear equation, there are no peaks or valleys.

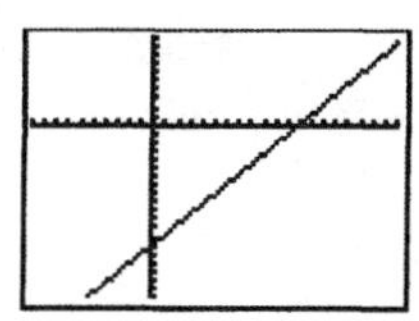

Reset the WINDOW to the standard viewing window.

Example 1: Graph $y = x^3 - 15x^2 + 26x$ in an appropriate viewing window.

Getting started	Enter $x^3 - 15x^2 + 26x$ at the Y1 prompt and view the graph in the standard viewing window. The y-intercept and one x-intercept are visible. An equation will have *at most* the same number of solutions as its degree. Therefore, in an nth degree equation you are looking for, at most, n x-intercepts.	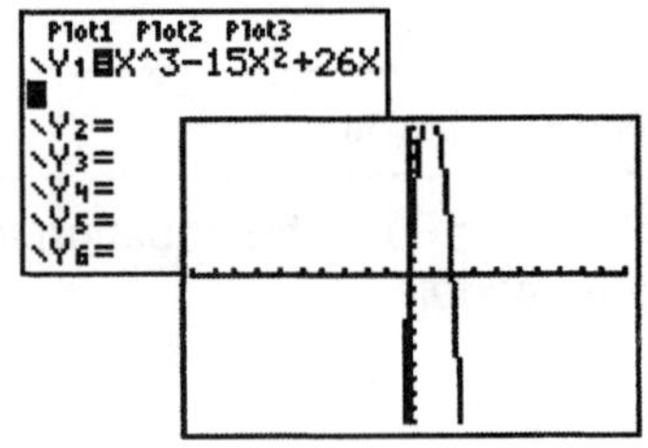
Find the *y*-intercepts	Press TRACE (or use the VALUE/EVAL feature) to find the y-intercept. y-intercept: (0,0)	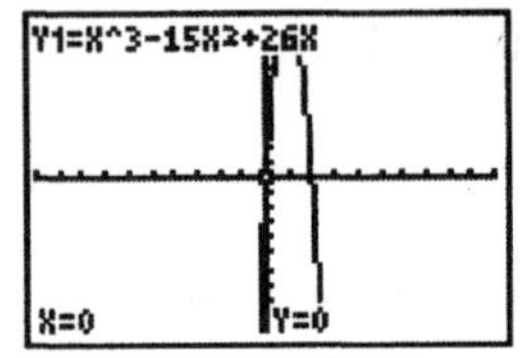
Find the *x*-intercept(s)	Two x-intercepts are visible, (0, 0) and (2, 0). We must find the third intercept. Points above an x-intercept will have positive y-values whereas points below an x-intercept will have negative y-values. Access the table and scroll through looking for zeros or changes in sign in the Y1 column. x-intercept(s): (0, 0), (2, 0), (13, 0)	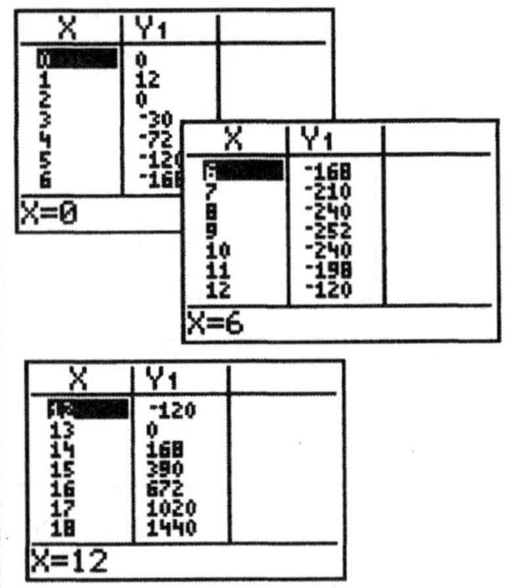
Reset the viewing window	The value of Xmax on the window screen is changed to 15 (a little above the x-intercept of 13). When we view the graph, the intercepts are clearly visible. We must now locate any high and/or low points of the graph.	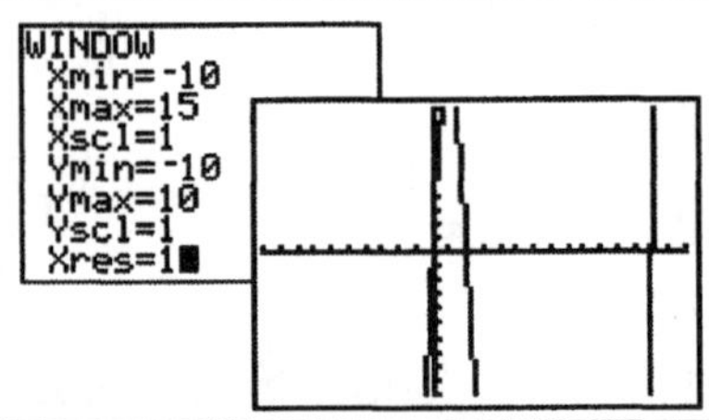
Locate high points and/or low points	Return to the table and locate the maximum/minimum y-values that occur between the x-intercepts. The largest relative y-value is 12 and the smallest relative y-value is - 252. Reset the viewing window, changing the value on Ymax to 15 and on Ymin to - 275. You may prefer to set the y-scale to a value besides 1, or to 0 to eliminate the tic marks entirely.	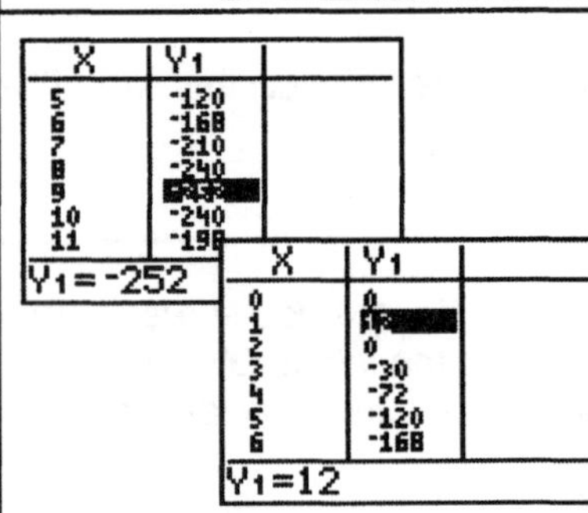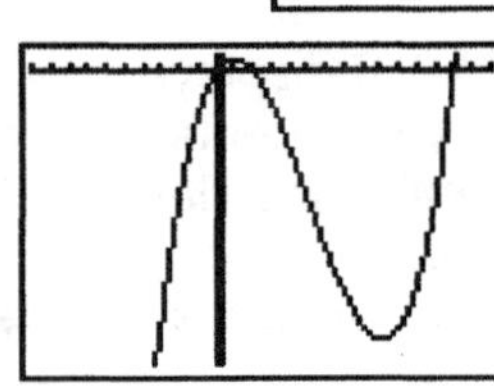

Procedure for Obtaining Good Viewing Windows	
1.	Begin by displaying the graph in the standard viewing window.
2.	Press TRACE or use the VALUE (EVAL) feature to display the coordinates of the y-intercept. If the intercept is out of the standard viewing window, adjust the window to accommodate the intercept and return to the graph screen.
3.	Access the TABLE feature (set it to begin at 0 and increment by 1) and scroll through the table looking for 0 or changes in sign between entries in the Y1 column . Remember, the polynomial will have *at most* the same number of x-intercepts as its degree.
4.	Reset the window using the values found for the intercepts (both x- and y-intercepts).
5.	To find high/low points of the graph scroll through the table and record maximum and minimum y-values between the x-intercepts.
6.	Reset the window using these values for Ymax and Ymin respectively.

As your algebraic knowledge increases, you will be able to blend this knowledge with calculator expertise to expedite the process of determining a good viewing window.

EXERCISE SET

Directions: Begin each problem by viewing the graph in the ZStandard viewing window.

a. Record the values used to determine your WINDOW.
b. Sketch the graph that is displayed for these WINDOW values.
c. Record the WINDOW in interval notation.

1. $y = -2x^2 + 19$

y-intercept = ______ Ymax = _____

x-intercepts = ______, and ______

[______, ______] by [______, ______]
Xmin Xmax Ymin Ymax

2. $y = ½x^4 - 10x^2 + 25$

y-intercept = _____

x-intercepts = ______, ______, ______, ______

Ymax = _____ Ymin = _____, _____

[______, ______] by [______, ______]
Xmin Xmax Ymin Ymax

3. $y = x^3 - 172x - 336$

y-intercept = _____

x-intercepts = ______, ______,

Ymin = _____

[______, ______] by [______, ______]
Xmin Xmax Ymin Ymax

4. Establish an appropriate viewing window for the equation $y = 0.1x^4 - 8x^2 + 5x + 1$ and sketch the result.

[______, ______] by [______, ______]

Solutions: **1.** *y*-intercept:19, Ymax: at least 20, *x*-intercepts: approximately -3 and 3; [-4, 4] by [-10, 20]

2. *y*-intercept:25, *x*-intercepts: approximately -4, -1.7, 1.7, 4, Ymax:25, Ymin:-26, -26; [-5, 5] by [-30, 30]

3. *y*-intercept: - 336, *x*-intercepts: - 2, - 12, 14, Ymin: -1200; Ymax: 528 [-25, 20] by [-1500, 600]

4. [-10, 10] by [-200, 10]

UNIT 15
CALCULATOR VIEWING WINDOWS

*Prerequisite: Unit #3

TI-86 IF USING THE TI-86, GO TO THE GUIDELINES (PG.95).

Setting Up The Graphical Display

The calculator's display is controlled through the **MODE** and **FORMAT** screens.

Press the **[MODE]** key. **MODE** controls how numbers and graphs are displayed and interpreted. The current settings on each row should be highlighted as displayed. The blinking rectangle can be moved using the four **cursor** (arrow) keys. To change the setting on a particular row, move the blinking rectangle to the desired setting and press **[ENTER]**.

NOTE: Items must be highlighted to be activated.

Normal vs. Scientific notation
Floating decimal vs. Fixed to 9 places
Type of angle measurement
Type of graphing: function, parametric, polar, sequence
Graphed points connected or dotted
Functions graphed one by one
Numbers can be viewed as as Real or complex
Screen can be split to view two screens simultaneously

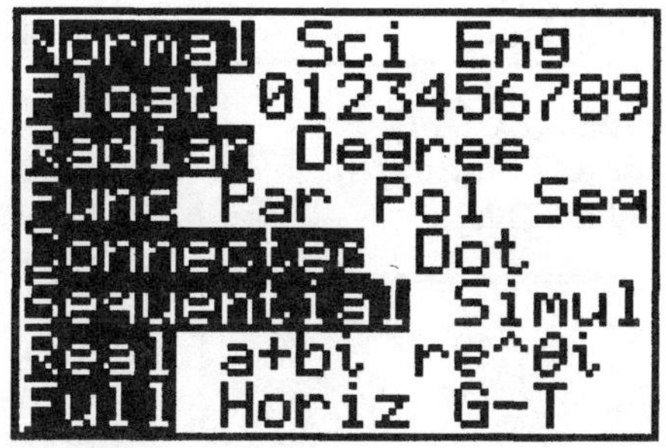

To return to the home screen, at this point, press **[CLEAR]** or **[2nd] <QUIT>**. The FORMAT screen controls the way graphs are displayed. Access this screen by pressing **[2nd] <FORMAT>** (located above the ZOOM key directly under the graph screen). Each item on the left should be hightlighted as pictured. Each menu option is briefly described below.

Graphs on a rectangular coordinate system
Cursor location is displayed on screen
Graphing grid is not displayed
Axes are visible
Axes are not labeled with an X and Y
The expression entered at Y1 is displayed on the graph when trace is activated.

Press **[Y=]**. The calculator can graph up to 10 different equations at the same time. Because **MODE** is in the sequential setting, the graphs will be displayed sequentially. Note that cursoring down accesses additional Y= prompts. Clear any entries at any of the Y= prompts. The display of the equations entered on the **Y=** screen is controlled by the size of the viewing window. The dimensions of the viewing window are determined by the values entered on the **WINDOW** screen.

The TI-83/84 series calculators have a feature called the *graph style icon* that allows you to distinguish between the graphs of equations. To change styles press **[Y=]** and observe the "\" in front of each Y. Use the left arrow to cursor over to this "\." The diagonal should now be moving up and down. Pressing

[ENTER] once changes the diagonal from thin to thick. The graph of Y1 will now be displayed as a thick line. Repeatedly pressing **[ENTER]** displays the following:

◥: shades above Y1
◣: shades below Y1
-o: traces the leading edge of the graph followed by the graph
o: traces the path but does not plot
⋱ : displays graph in dot, not connected MODE

It is important to remember that the graphing icon takes precedence over the **MODE** screen. If the icon is set for a solid line and the **MODE** screen is set for DOT and not solid, the graphing icon will determine how the graph is displayed. Pressing **[CLEAR]** to delete an entry at the Y= prompt will automatically reset the graphing icon to default, a solid line.

Standard Viewing Window	**Interpretation**	
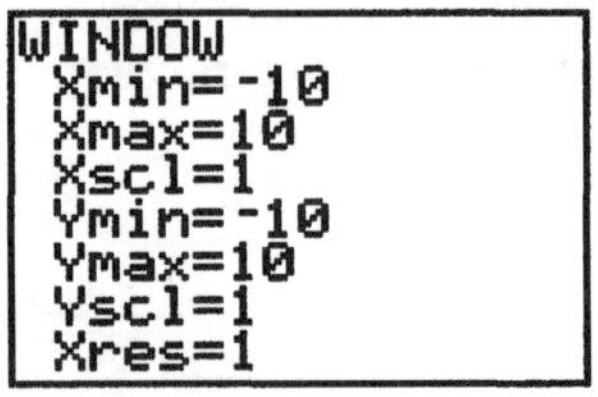 Press **[ZOOM]** **[6] (6:ZStandard)** **[WINDOW].** 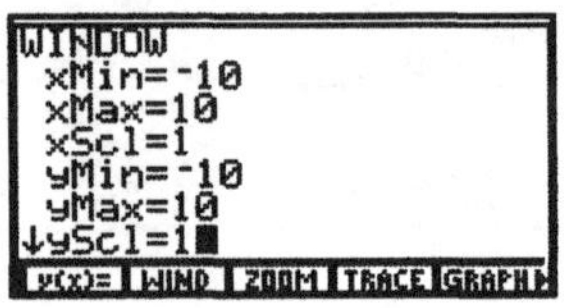 Press **[GRAPH]** **[F3](ZOOM) [F4](ZSTD)** **[2nd] [F2](WIND).**	Xmin: the smallest value of *x* Xmax: the largest value of *x* Xscl: the number of units per tic mark Ymin: the smallest value of *y* Ymax: the largest value of *y* Yscl: the number of units per tic mark The values entered on this screen may be changed by using the cursor arrows to move to the desired line and typing over the existing entry. The row labeled Xres=determines the screen resolution. It should be set equal to 1, which means that each pixel on the X-axis will be evaluated and graphed.	10 -10 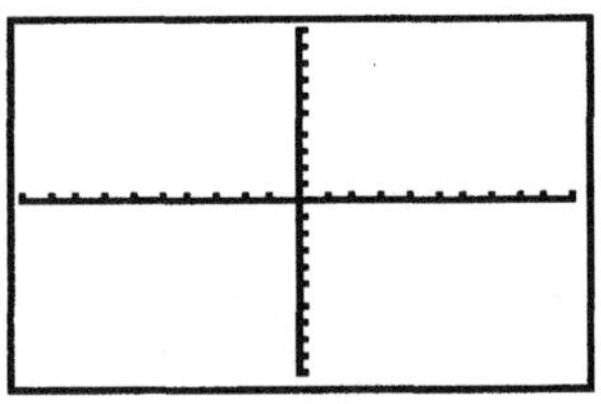 10 -10 This is called the standard viewing window. The interval notation for this information is [-10,10] by [-10,10].

EXERCISE SET

Directions: Before proceeding further, press [Y=] and clear all entries.

1. Press **[WINDOW]** and enter Xmin = -5, Xmax = 5, Xscl = 1, Ymin = -12, Ymax = 7, Yscl = 1. Be sure to use the gray **[(-)]** key for negative signs. Press **[GRAPH]** to view the coordinate axes. Count the tic marks on the axes and see how these marks correspond to the max and min values. Label the last tic mark on each axis (i.e. farthest tic mark left, right, up and down) with the appropriate integer value.

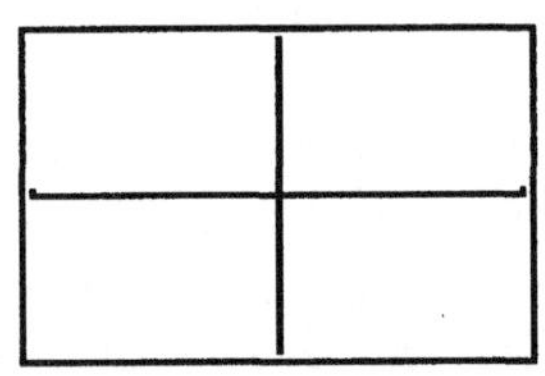

TI-86	TI-86 USERS MUST PRESS **[CLEAR]** TO DELETE MENU DISPLAY BEFORE COUNTING TIC MARKS.

✍2. Change the viewing window to Xmin = -20, Xmax = 70, Xscl = 10, Ymin = -5, Ymax = 15, Yscl = 3 and press **[GRAPH]** to view the axes. How many tic marks are on the positive portion of the *x*-axis?____ How many units does each of the tic marks on this axis represent?____ Based on your last two answers, how many units long is the positive portion of the *x*-axis?____ Does this number correspond to the Xmax value given in the problem? ________

In your own words, explain what is happening.

✍3. To help "de-bug" errors in graphing set ups later on, describe what you think would happen if Xmin = 10 and Xmax = -5. You might want to draw your own set of coordinate axes and try to label them in this manner. Enter these values on the **WINDOW** screen and press **[GRAPH]**. What did happen?

✍4. Reset Xmin = -10, Xmax = 10 and describe what you think will happen if you set Ymin = 5, Ymax = 5. Again, enter the values and press **[GRAPH]**. What did happen? (Try drawing your own set of axes and labeling them as indicated.)

✍5. What should the relationship between Max and Min be? (i.e. Min > Max, Min < Max, or Min = Max)

Entering Expressions to be Graphed: The [Y=] Key

Reset the viewing window to ZStandard, by pressing **[ZOOM] [6:ZStandard]** (TI-86 users press **[GRAPH] [F3](ZOOM) [F4](ZSTD)**).

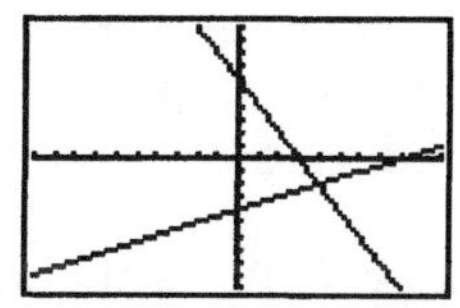

Press **[Y=]** (TI-86 users: because *y*(*x*)= is on the second line of graph menu, press **[2nd]<M1>(*y*(*x*))** . On the screen Y1= is followed by a blinking cursor. Anything else can be cleared by pressing **[CLEAR]**. Enter -2*x*+ 6 at the Y1 prompt and press **[ENTER]**. The cursor is now on the second line following Y2=. At this prompt, enter the expression (1/2)*x* - 4. Note that the equal signs beside both Y1 and Y2 are highlighted. This means that both equations will be graphed. Press **[GRAPH]** (TI-86: because GRAPH is located on the second line of menu display, press **[2nd]<M5>(GRAPH)**) to display the graph screen.

Graphing

To graph $y = -2x + 6$ only, press **[Y=]** and use the arrow key to move the cursor over the equal sign beside Y2. Press **[ENTER]**. Notice that the equal sign beside Y2 is *not* highlighted, whereas the equal sign beside Y1 *is* highlighted. Press **[GRAPH]**; only the highlighted equation, Y1, is graphed.

TI-86	TO GRAPH Y = -2X + 6 ONLY, PRESS **[F1](Y(X)=)**, PLACE THE CURSOR ON THE Y2 EQUATION AND PRESS **[F5](SELCT)**. THE EQUAL SIGN IS NO LONGER HIGHLIGHTED, INDICATING THAT THE GRAPH OF Y2 WILL NOT BE DISPLAYED. THE Y2 EQUATION CAN BE RESELECTED FOR GRAPHING BY PLACING THE CURSOR ON THE EQUATION AND PRESSING **[F5](SELCT)** AGAIN.

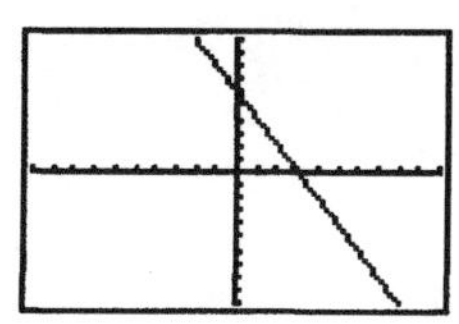

On the viewing screen at right, the calculator draws a set of axes whose minimum and maximum values and scale match the choices under **WINDOW**. The graph of Y1 is drawn from left to right. Return to the Y= menu by pressing **[Y=]**. Cursor down to Y2 and *turn on* this graph by highlighting the equal sign. Press **[GRAPH]** and notice that the two graphs are drawn in sequence. **SEQUENTIAL** was chosen from the **MODE** menu earlier.

TI-86	TI-86 USERS GO TO THE GUIDELINES (PG.96), AND READ THE SECTION ENTITLED ALTERING THE VIEWING WINDOW.

Zoom Menus

Because the screen is 95 pixel points wide by 63 pixel points high, there are 94 horizontal spaces and 62 vertical spaces to light up. When tracing on a graph, the readout changes according to the size of the space. The size of the space can be controlled by the following formulas:

$$\frac{Xmax - Xmin}{94} = \text{horizontal space width}, \quad \frac{Ymax - Ymin}{62} = \text{vertical space height.}$$

We will examine some preset viewing windows and how they affect the pixel space size.

Press **[ZOOM]**. There are ten entries on this screen. The down arrow key can be used to view remaining entries.

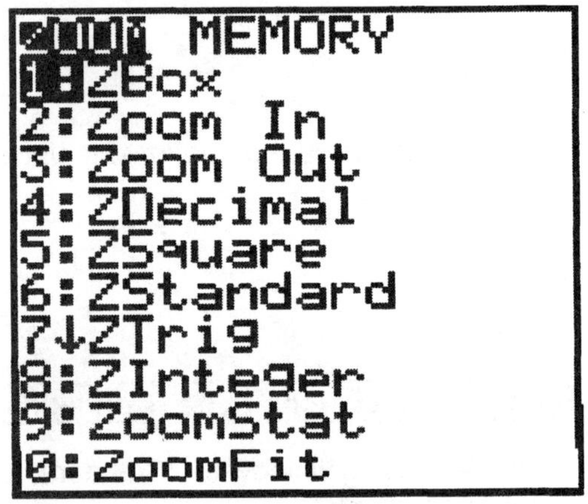

1: Boxes in and enlarges a designated area.
2: Acts like a telephoto lens and "zooms in."
3: Acts like a wide-angle lens and "zooms out."
4: Cursor moves are ONE tenth of a unit per move.
5: "Squares up" the previously used viewing window.
6: Sets axes to [-10,10] by [-10,10].
7: Used for graphing trigonometric functions.
8: Cursor moves are ONE integer unit per move.
9: Used when graphing statistics.
0: Replots function, recalculating Ymin and Ymax.

We will now look at the graphs in some of the preset windows that we access from the ZOOM menu. Clear your current entries at the Y1 and Y2 prompts on your *y*-edit screen. Enter -2*x* + 3 at the Y1 prompt and enter ½ *x* -2 at the Y2 prompt. Look at each graph in the specified screen and TRACE along the graph, observing the *x* and *y*-coordinates displayed at the bottom of the screen.

Note: TI-86 users press **[CLEAR]** after the graph is displayed to remove the menu line. Your graph should match the one pictured.

Menu Option/Keystrokes	Window Values/Graphical Display/Notes
ZDecimal **[ZOOM] [4] (4:ZDecimal)** TI-86: with one menu line displayed , **[F3] (ZOOM) [MORE] [F4] (ZDECM)** If one menu line is *not* displayed, press **[GRAPH]** followed by the specified keystrokes.	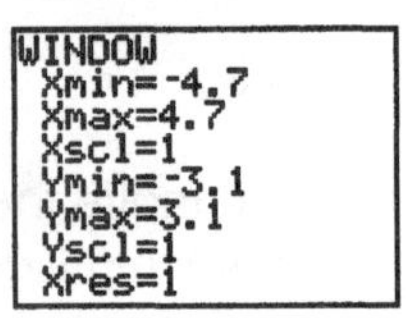 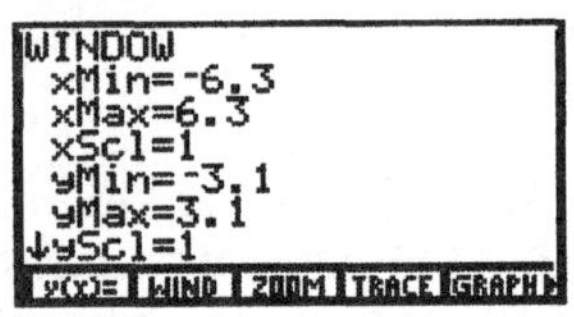 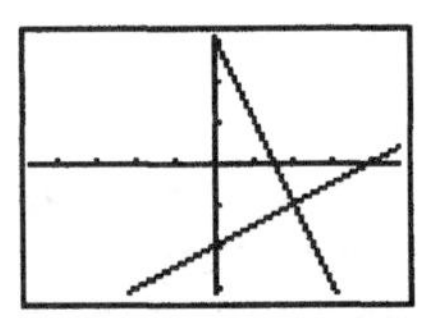 **ZDecimal** is useful for graphs that require the use of the calculator's TRACE feature. Applying the horizontal space width formula, $\frac{\text{Xmax-Xmin}}{94} = \frac{4.7-(-4.7)}{94} = 0.1$, changes the *x*-values by one-tenth of a unit each time the cursor is moved. In general, Xmax - Xmin needs to be a multiple of 94 to produce friendly trace values. TI-86 users should observe the formula differences: $\frac{\text{Xmax-Xmin}}{126} = \frac{6.3-(-6.3)}{126} = 0.1$

ZInteger Press **[ZOOM] [8] (8:ZInteger)**, pause for the graph to be displayed, and move the cursor to the origin of the graph (x = 0 and y= 0). Press **[ENTER]** . TI-86: Press **[GRAPH] [ZOOM] [MORE] [MORE] (ZINT)** and move the cursor as directed above.	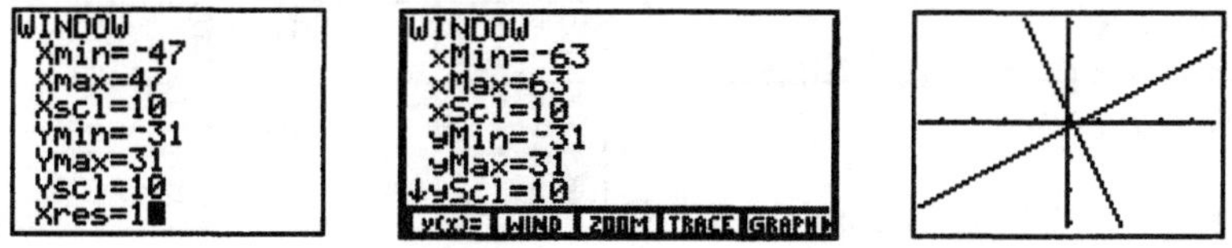 **ZInteger** is useful for application problems where the *x*-value is valid only if represented as an integer (such as when *x* equals the number of tickets sold, number of passengers in a vehicle, etc.). The horizontal space width formula, $\frac{\text{Xmax - Xmin}}{\text{94 or 126}} = 1$, changes the *x*-values by one unit each time the cursor is moved.
ZStandard Press **[ZOOM]** **[6] (6:ZStandard)**. TI-86: Press **[GRAPH]** **[F3] (ZOOM) [F4] (ZSTD)**.	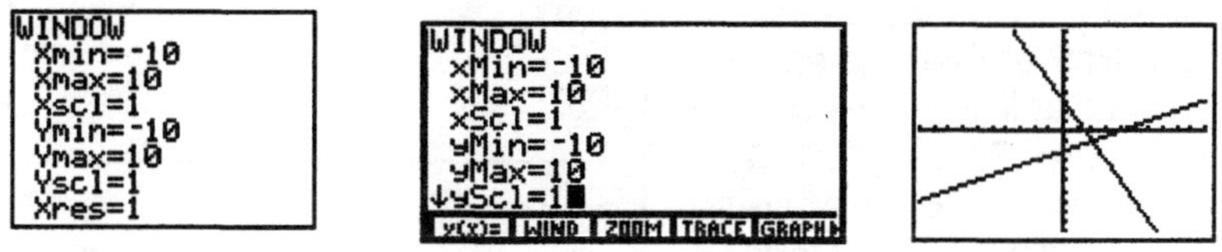 **ZStandard** provides a good visual comparison between hand sketched graphs (or textbook graphs) that are approximately [-10,10] by [-10,10]. Applying the horizontal space width formula, $\frac{\text{Xmax - Xmin}}{\text{94 or 126}} = 0.212765974$, the *x*-values will change by .212765974 each time the TRACE cursor is moved.
ZDecimal x *n* Access WINDOW for the ZDecimal screen. Multiply the Xmin and Xmax by the same constant and the Ymin and Ymax by the same constant (*n*) to produce a larger viewing rectangle which still provides cursor moves in tenths of units.	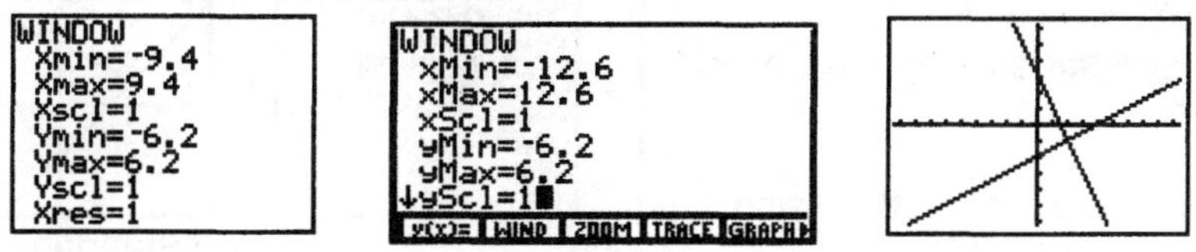 **ZDecimal** frequently does not provide a large enough viewing window. Multiplying Xmin and Xmax by 2 would mean cursor moves of two-tenths of a unit: $\frac{\text{Xmax - Xmin}}{\text{94 or 126}} = 0.2$, whereas multiplying by 3 would mean cursor moves of three-tenths of a unit. The screen above is the ZDecimal screen with the max and min values multiplied by 2.
Zoom In Display the graph in a ZStandard viewing window. Press **[ZOOM] [2] (Zoom In)**. View the graph. Pressing **[ENTER]** again, zooms in more. TI-86: Press **[F3] (ZOOM) [F2] (ZIN)[ENTER]**.	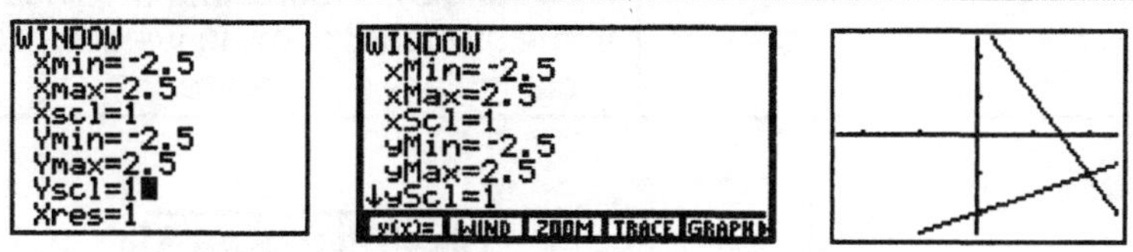 Once you've zoomed in, pressing **[ENTER]** will zoom in on the graph once again.

Zoom Out Display the graph in a ZStandard viewing window. Press **[ZOOM] [3] (Zoom Out) [ENTER].** TI-86: Press **[F3] (ZOOM) [F3] (ZOUT) [ENTER]**	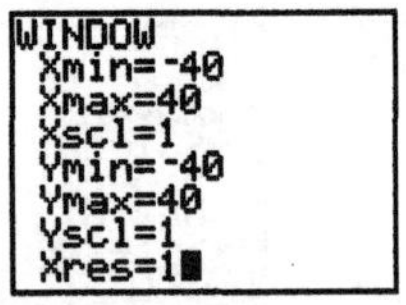 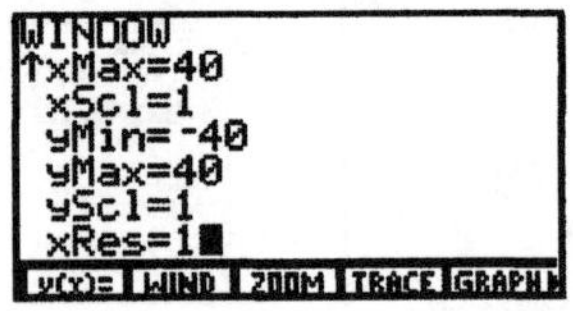 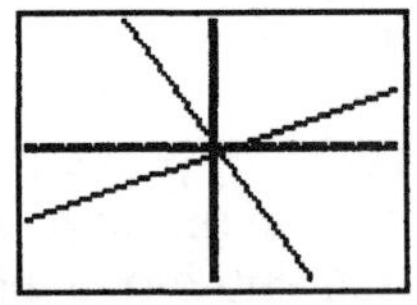 Notice the distortion of the *x*- and *y*-axes. This is because the WINDOW screen has specified a scale of one unit on both axes. Experiment with using different values for the *x* and/or *y*-scale. Once you've zoomed out, pressing **[ENTER]** will zoom out on the graph once again.
ZBox To magnify a specific section of a graph (rather than the entire graphical display) we will use the ZBox feature. Press **[ZOOM] [1] (1:ZBox)** (TI-86 users press **[ZOOM] [F1](BOX)**). Move cursor to upper left corner of desired box, press **[ENTER]**, move cursor to upper right corner of desired box and down to lower right corner. Press **[ENTER]** to enlarge boxed area.	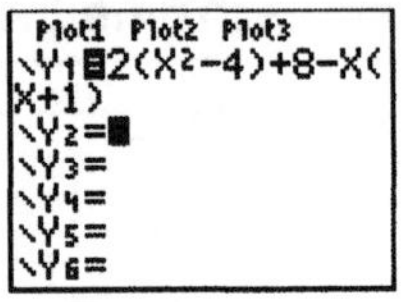 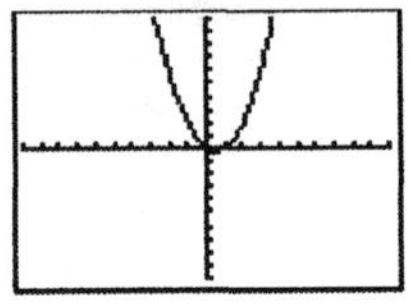 We cannot determine if the graph simply touches the x-axis at a single point or if it dips below the axis and intersects it at two points. 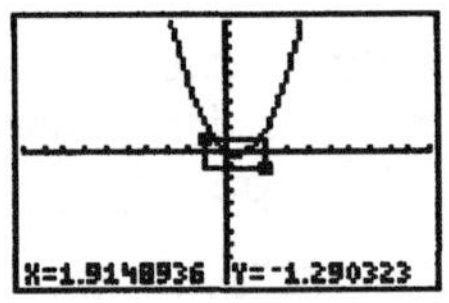 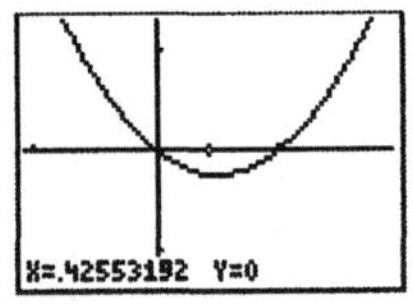 ZBox
ZSquare ZSquare *squares up* the viewing screen. Begin in the standard viewing window. Press **[ZOOM] [5] (5:ZSquare)**. TI-86: Press **[F3] (ZOOM) [MORE] [F2] (ZSQR)**.	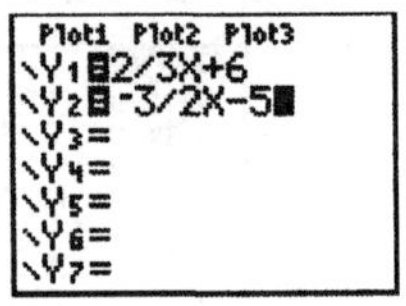 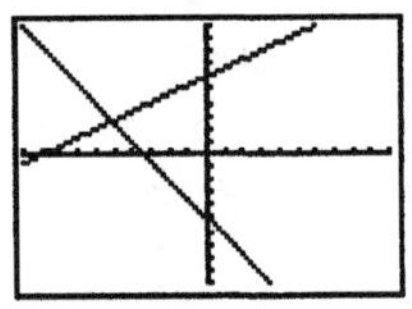 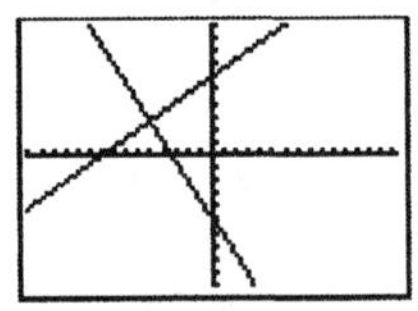 ZStandard Press **[Y=]** and clear all entries. Enter Y1 $=\frac{2}{3}x+6$ and Y2$=-\frac{3}{2}x-5$. These equations produce perpendicular lines (their intersection forms a 90° angle). Notice that the lines do not appear to be perpendicular in the standard viewing window. This is because the screen is rectangular - not square.

EXERCISE SET

✍6. Press **[WINDOW]** to see how the Max and Min values were affected by the Zsquare command applied above. Enter the WINDOW values displayed. Explain how the viewing window is different from the ZStandard viewing window.

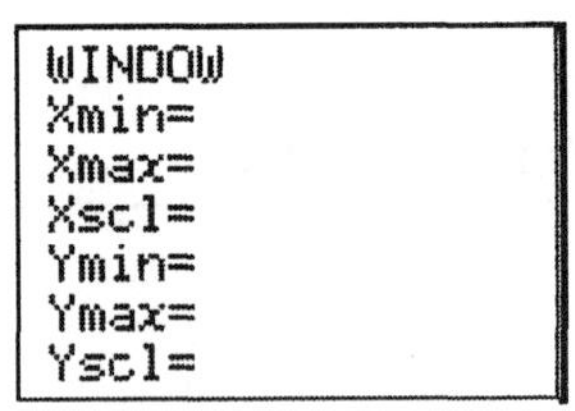

NOTE: The ZDecimal screen (or any multiplicity of this screen) will provide a "squared up" graph screen on the TI-83/84 series. TI-86 users will need to use ZSquare .

7. **CLEAR** all entries on the **Y=** screen. Graph Y1 $= \frac{1}{2}x - 4$ in the ZDecimal viewing window. Sketch the screen. Write the appropriate value for the "endpoint" of each axis on the graph.

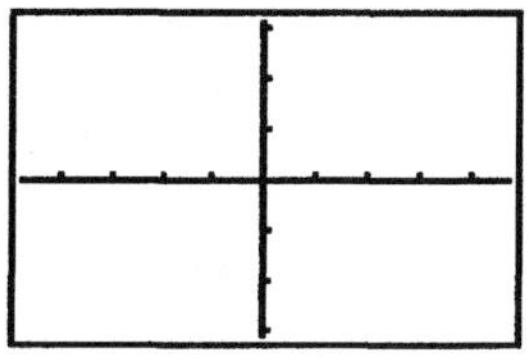

8. Now, access Zinteger. Sketch the screen as displayed. Write the appropriate value for the "endpoint" of each axis on the graph.

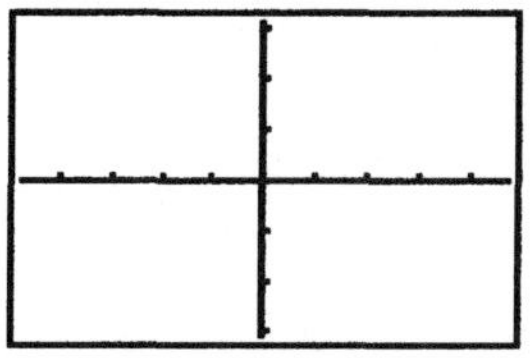

NOTE: ZDecimal and ZStandard do not require you to press [ENTER] to activate the viewing WINDOW, but ZInteger demands it.

9. In your own words, explain the differences in the displays in #7 and #8. What accounts for these differences?

10. Enter the equation $y = 12x^6 - 58x^4 + 84x^2 + 8$ at the **Y1** prompt. Try to view the graph in each of the pre-set viewing windows discussed in this unit - ZDecimal, ZStandard, ZInteger, then sketch the graph as displayed in each of the indicated viewing WINDOWS. Both the Xscl and Yscl should be equal to zero.

[0,2] by [15,35] [-2,2] by [-5,30] [-5,5] by [-5,75]

Pre-set viewing WINDOWS can provide a "starting point" for displaying a complete graph but frequently do not display all of the critical features of the graph.

Solutions: **1.** left: -5, right: 5, top: 7, bottom: -12 **2** 7, 10, 70, yes; By increasing the value of the scale, the axes are increased in size without physically extending the length.

3. Because the maximum does not exceed the minimum (i.e. -5 is NOT greater than 10) the calculator displayed an error message. Moreover, that error message tells you that <u>you</u> made an error in setting the values.

4. The y-axis has not been given a defined length by setting the max and min at the same value. Again, an error message is displayed.

5. Max > Min **6.** To have a "square" screen, tic marks must be evenly spaced. This was accomplished by adding tic marks to the *x*-axis.

7. 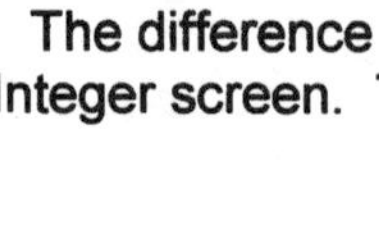

Xmin: -4.7, Xmax: 4.7, Ymin: -3.1, Ymax: 3.1
TI-86: xMin: -6.3, xMax: 6.3,
yMin: -3.1, yMax: 3.1

8. 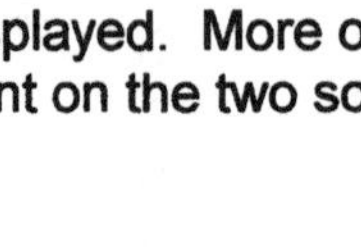

Xmin: -47, Xmax: 47, Ymin: -31, Ymax: 31
TI-86: xMin: -63, xMax: 63,
yMin: -31, yMax: 31

9. The difference was the amount and position of graph displayed. More of the graph was displayed on the ZInteger screen. This was because the <u>scales</u> were different on the two screens.

TI-86 GUIDELINES UNIT 15

Setting Up the Graphical Display

The calculator's display is controlled through the **MODE** and **GRAPH/FORMAT** screens.

Press **[2nd] <MODE>**. The current settings on each row should be highlighted as displayed. The blinking rectangle can be moved using the four **cursor** (arrow) keys. To change the setting on a particular row, move the blinking rectangle to the desired setting and press **[ENTER]**.

NOTE: Items must be highlighted to be activated.

Normal vs. Scientific notation
Floating decimal vs. Fixed to 11 places
Type of angle measurement
Complex number display
Type of graphing: function, polar, parametric, differential eq.
Performs computations in bases other than base 10
Format of vector display
Type of differentiation

The **FORMAT** screen is accessed by pressing **[GRAPH] [MORE] [F3](FORMT)**. The following settings should be highlighted. Each menu option is briefly described below.

Graphs on both rectangular and polar coordinate system
Cursor location is displayed on the screen
Graphed points are connected or discrete
Functions displayed sequentially or simultaneously
Graphing grid is not displayed
Axes are visible
Axes are not labeled with "x" and "y"

To return to the home screen, at this point, press **[CLEAR]** or **[EXIT]**.

Press **[GRAPH] [F1](y(x)=)**. The calculator can graph up to 99 different equations at the same time. Because **MODE** is in the sequential setting, the graphs will be displayed sequentially. The display of the equation(s) entered on the y(x)= screen is controlled by the size of the viewing window. The dimensions of the viewing window are determined by the values entered on the **WINDOW** screen (**WIND**).

A feature called the "graph style icon" allows you to distinguish between the graphs of equations. Press **[GRAPH] [F1](y(x)=)** and observe the "\" in front of each Y. Press **[MORE]** followed by **[F3](STYLE).** The diagonal should now have changed from thin to thick. The graph of Y1 will now be displayed as a thick line. Repeatedly pressing **[STYLE]** displays the following:

◥: shades above y1
◣: shades below y1
-o: traces the leading edge of the graph followed by the graph
o: traces the path but does not plot
⋱: displays graph in dot, not connected MODE

It is important to remember that the graphing icon takes precedence over the **MODE** screen. If the icon is set for a solid line and the **MODE** screen is set for DOT and not solid, the graphing icon will determine how the graph is displayed. Pressing **[CLEAR]** to delete an entry at the Y= prompt will automatically reset the graphing icon to default, a solid line.

☞ RETURN TO THE CORE UNIT (PAGE 88) BEGIN READING AT THE BOXED INFORMATION "STANDARD VIEWING WINDOW"

Zoom Menus

Because the screen is 127 pixel points wide by 63 pixel points high, there are 126 horizontal spaces and 62 vertical spaces to light up. When tracing on a graph, the readout changes according to the size of the space. The size of the space can be controlled by the following formulas:

$\frac{Xmax - Xmin}{126}$ = horizontal space width, $\frac{Ymax - Ymin}{62}$ = vertical space height. We will now examine some preset viewing windows and how they affect the pixel space size.

Press **[GRAPH]** **[F3](ZOOM)**. Pressing **[MORE]** repeatedly will display the remaining menu selections and then return the display to the original screen.

BOX	Boxes in and enlarges a designated area.
ZIN	Acts like a telephoto lens and "zooms in".
ZOUT	Acts like a wide-angle lens and "zooms out".
ZSTD	Automatically sets standard viewing window to [-10,10] by [-10,10].
ZPREV	Resets **RANGE** values to values used prior to the previous ZOOM operation.
ZFIT	Resets yMin and yMax on the **RANGE** screen to include the minimum and maximum y-values that occur between the current xMin and xMax settings.
ZSQR	"Squares up" the previously used viewing window.
ZTRIG	Sets the **RANGE** to built-in trig values.
ZDECM	Sets cursor moves to ONE tenth of a unit per move.
ZDATA	*TI-86 only: Automatically sets the viewing window to accomodate statistical data.*
ZRCL	Sets **RANGE** values to those stored by the user (see ZSTO).
ZFACT	Sets the zoom factors used in **ZIN** and **ZOUT**.
ZOOMX	Graph display is based on xFact only when zooming in or out.
ZOOMY	Graph display is based on yFact only when zooming in or out.
ZINT	Cursor moves are ONE integer unit per move.
ZSTO	Stores current **RANGE** values for future use. Values are recalled by
ZRCL.	

☞ RETURN TO CORE UNIT PG.90 AND BEGIN READING BELOW THE LARGE TI-83/84 SERIES SCREEN.

UNIT 16
FUNCTIONS

*Prerequisite: Unit #15

A function is a relation that pairs each input value (x) with exactly one output value, (y).

Domain and Range

The set of all input values, x, is the domain; the set of all output values, y, is the range. When determining the domain and range of a function there are three tools available to you: the algebraic definition of the function, the table feature of the graphing calculator, and the graph. When reading a graphical display, domain is read from left to right and range is examined from low to high. We are only graphing real values and care must be taken when working with radical and rational functions.

Algebraically	**Numerically**	**Graphically**	**Conclusion**
$f(x) = 3x - 4$ The input value can be any real number.	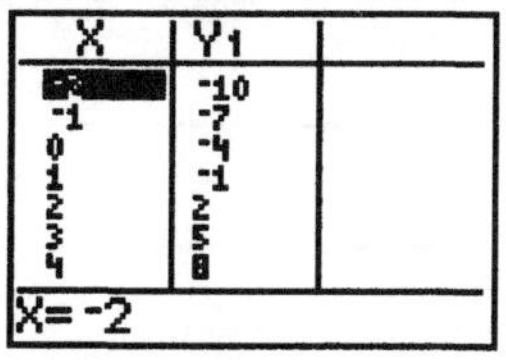 Scrolling through the table supports our algebraic conjecture.	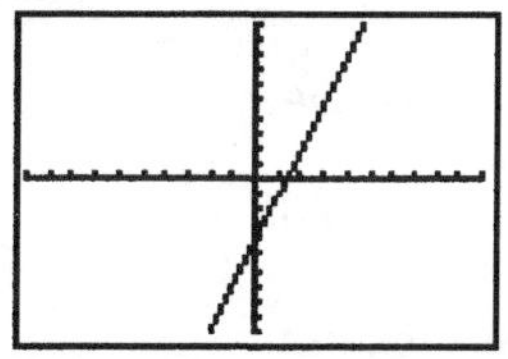 The graph supports our conjecture about domain; the range is all real numbers.	Domain: $\mathbb{R}$ Range: $\mathbb{R}$
$f(x) = \|x + 2\|$ The input value can be any real number. The output value would be greater than or equal to 0.	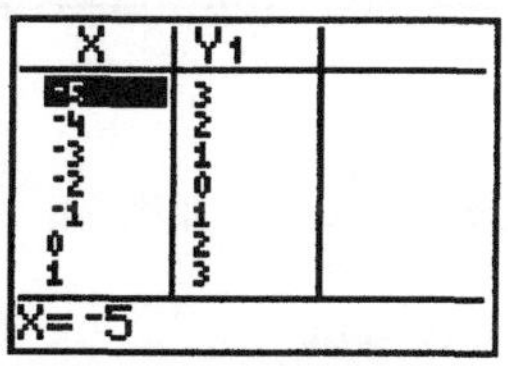 Scrolling through the table supports our algebraic conjecture about domain and range.	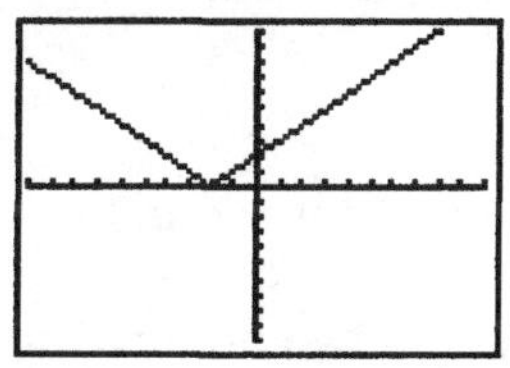 The graph supports our conclusions.	Domain: $\mathbb{R}$ Range: $\{y \mid y \geq 0\}$ Interval notation: $[0, \infty)$
$f(x) = \sqrt{x+3}$ The input value must be -3 or greater to be defined in the real number system. The output values would greater than or equal to 0.	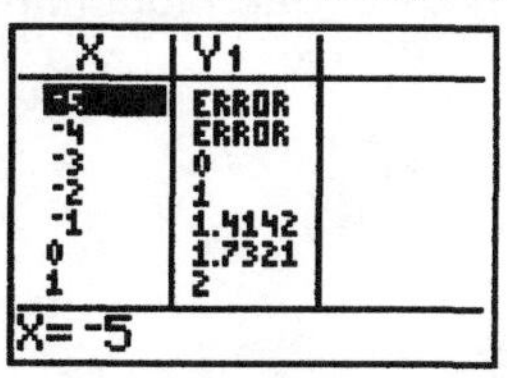 The ERROR message for input values smaller than -3 supports our conclusion that x must be greater than or equal to -3. The range appears to be positive or 0.	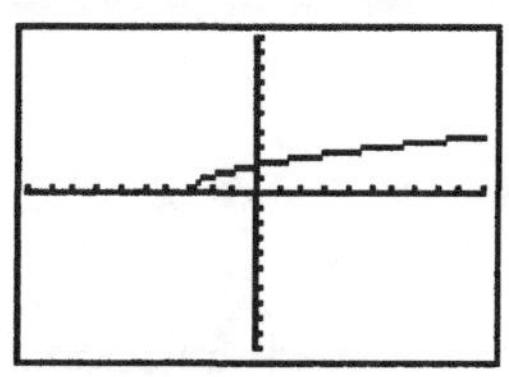 The graphical display supports our previous conclusions	Domain: $\{x \mid x \geq -3)$ Interval Notation: $[-3, \infty)$ Range: $\{y \mid y \geq 0\}$ Interval notation: $[0, \infty)$

It is suggested that you scroll up and down the table and trace along the graph before coming to a conclusion about domain and range.

EXERCISE SET

Directions: Use the **TRACE** feature to determine the domain and range of each of the following functions. TRACE along the path of the graph and examine the *x* and *y* values displayed at the bottom of the screen. These will be of assistance in determining the domain and range. You may use any viewing window you desire, however you may discover that some viewing windows are more informative than others. Suggestion: view and TRACE on each graph in each of the following windows - ZStandard, ZDecimal, ZInteger before determining the domain. (TI-86 users recall that these preset windows are listed under the **ZOOM** menu as ZSTD, ZDECM, and ZINT.)

1. $y = x^2 + 1$ Domain = ________________

   Range= ________________

2. $y = 3x + 5$ Domain = ________________

   Range = ________________

3. $y = \sqrt{x+5}$ Domain = ________________

   Range = ________________

4. $y = x^3 + 4x^2 + 2$ Domain = ________________

   Range = ________________

5. $y = \dfrac{x+2}{x}$ Domain = ________________

✍ 6. Consider the two functions $y = x$ and $y = \sqrt{x} \cdot \sqrt{x}$ and respond to each of the following questions.

   a. Would you expect the two domains to be the same? Why or why not?

   b. Considering only the algebraic equations, what is the domain of each function?

   c. What does the table of each function indicate as domain?

   d. What does the graph of each function indicate the domain should be? Is one viewing window more informative than another?

7. Consider the function $y = \frac{(2x+3)(x+2)}{2x+3}$ and respond to each of the following questions.

a. What do you expect the domain to be?

b. Considering only the algebraic equation, what is the domain?

c. What does the table of the function indicate as the domain?

d. What does the graph of the function indicate the domain should be? Is one viewing window more informative than another?

Inverses

The inverse of a function takes *f*(*x*) (the value of *y*) as its input value and *x* as the output value.

Function	Numerically/Graphically	Inverse	Numerically/Graphically
f(*x*) = 2*x* + 3	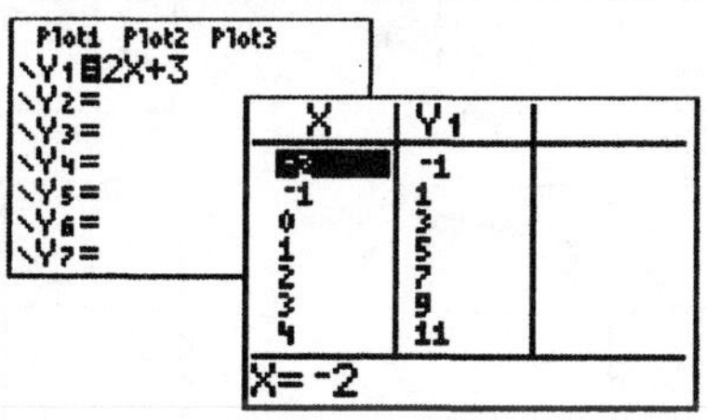 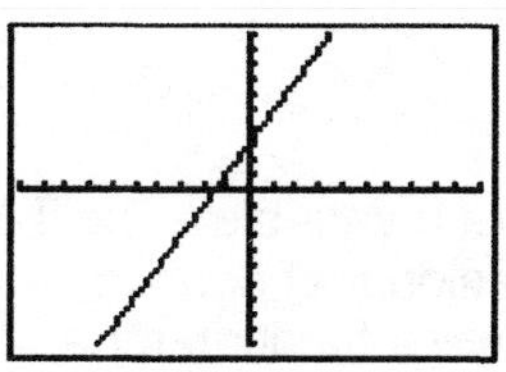	$f(x) = \frac{1}{2}x - \frac{3}{2}$	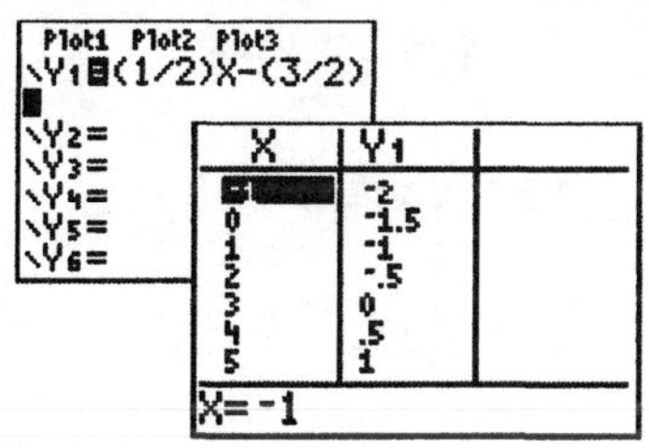 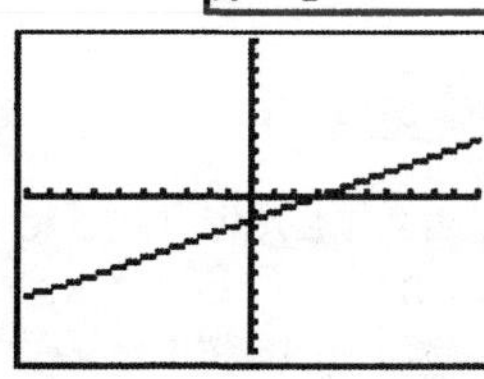
Observations	The function contains the ordered pairs (-2, -1), (-1, 1), (0, 3), and (5, 1) whereas the inverse contains the ordered pairs (-1, -2), (1, -1), (3, 0) and (5, 1). If the function and its inverse are graphed on the same set of coordinate axes along with *y* = *x*,(graphed as a thick line) the function and its inverse are symmetric about *y* = *x*.		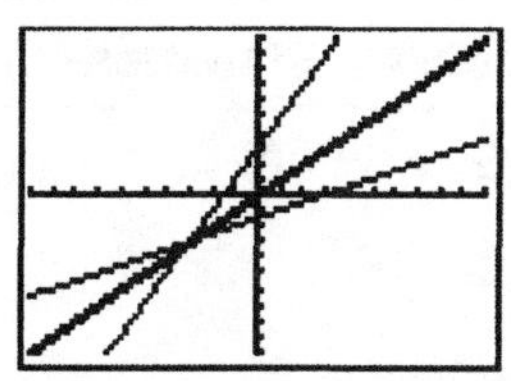

The calculator has a draw menu which will draw the inverse on the graph screen. To use the calculator to **DRAW** the inverse function, the function must be entered on the **Y=** screen. You cannot interact on a graph that is *drawn* on the screen rather than *graphed.* That is to say, you will not be able to TRACE, use INTERSECT, ROOT/ZERO, VALUE, TABLE, etc. on the graph of the inverse.

Function	Calculator Screens		
$f(x) = 2x + 3$	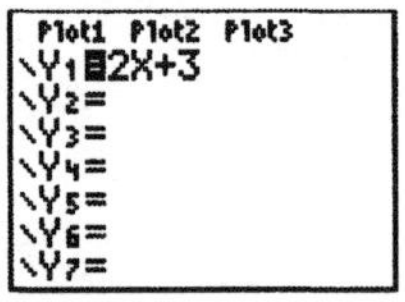 Enter the function at the Y1 prompt.	 From the home screen, instruct the calculator to **DRAW** the inverse of Y1 (DrawInv): Press **[2nd] <DRAW> [8] (8:DrawInv) [Vars] [▸]**(to highlight Y-Vars) **[1] (1:Function...) [1] (1:Y1) [ENTER].** TI-86: The draw inverse command is accessed from either the home screen or the graph screen by pressing **[GRAPH] [MORE] [F2](DRAW)** and pressing **[MORE]** until **(DrInv)** is displayed. Press the appropriate **F** key. Displayed on the home screen is the **DrInv** command. Enter y1 after the command by pressing **[2nd] [ALPHA] <y> [1] [ENTER].** The variable **y** ***must*** be lower case.	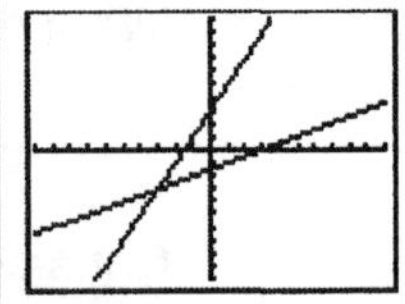 Screen display *after* pressing **[ENTER]**.

EXERCISE SET CONTINUED

Directions: Match each function to the graph of its inverse.

_______ 8. $f(x) = 3x + 4$

_______ 9. $f(x) = 4x - 8$

_______ 10. $f(x) = x^3 - 8$

_______ 11. $f(x) = (x-2)^2,\ x \geq 2$

A.
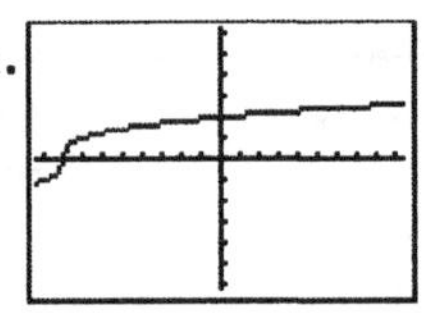

B.
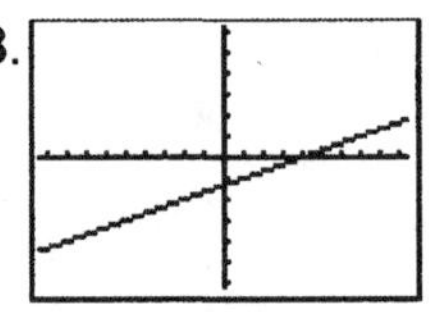

C.
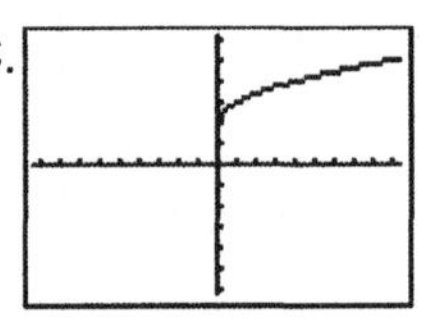

D.
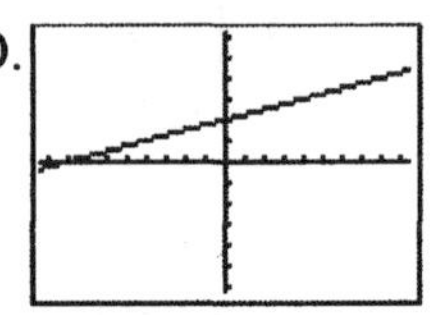

Evaluating Functions

The STOre feature may be used to evaluate a function for any value of a variable. The TABLE and VALUE features (EVAL on the TI-86) may also be used to evaluate an expression that contains the variable *x*. The calculator has another feature, VARS, that can also be used to evaluate a function for a specified value of *x*.

Algebraic Statement	VARS Feature	
If $f(x) = x^2 - 8x - 10$, find $f(2)$.	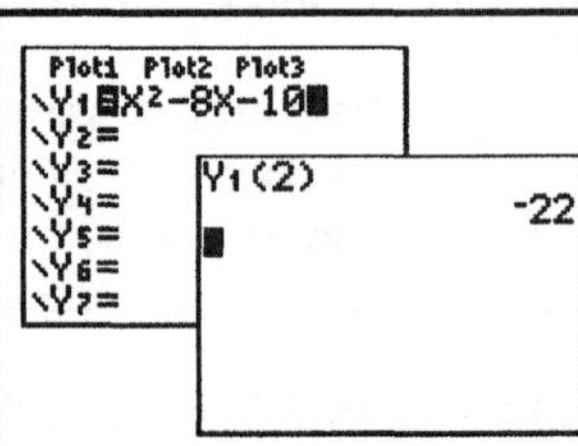 As an ordered pair this would be written as (2, -22).	Enter the polynomial at the Y1 prompt. Return to the home screen by pressing **[2nd]<QUIT>**. Press **[VARS] [▸] [1] (1:function) [1] (1:Y1) [( ] [2] [ )] [ENTER]**. TI-86 users press **[2nd] <alpha> <y> [1] [(] [2] [)]**.

EXERCISE SET CONTINUED

Directions: Enter each of the following functions on the **Y=** screen. Evaluate for the indicated value of *x* using the VARS capability of the calculator. Copy the screen that displays the answer. Record your final information as an ordered pair.

12. Evaluate $y = 3x^3 - 2x^2 + x - 5$ for $x = -\frac{3}{20}$.

Screen display:

Ordered pair:____________________________ (in decimal form)

Ordered pair:____________________________ (in fraction form)

13. Evaluate $y = \sqrt{2x-5}$ for $x = 3.5$.

Screen display:

Ordered pair:__________________

14. Evaluate $y = \dfrac{2x+3}{x^2+4x-5}$ for $x = -\frac{1}{2}$.

Screen display:

Ordered pair:__________________ (in fraction form)

✍15. Evaluate $y = \sqrt{3x+5}$ for $x = -8$.

Screen display:

Why did you get this display? Explain carefully.

16. The profit or loss for a publishing company on a textbook supplement can be represented by the function $f(x) = 10x - 15000$ (x is the number of supplements sold and $f(x)$ is the resulting profit or loss).

a. Use the VARS feature to determine the amount of profit (or loss) if 2000 supplements are sold. Copy the screen display to justify your work.

✍ b. How can you tell if the $5000 is profit or loss?

c. What would be the profit (or loss) if 1000 supplements are sold? Copy the screen display to justify your work.

✍17. In your own words, explain the various approaches for determining domain and range of functions. Include both algebraic and calculator methods in your discussion.

Solutions: **1.** domain= ℝ, range = $\{y|y\geq 1\}$ **2.** domain =ℝ, range = ℝ

3. domain = $\{x|x \geq -5\}$, range = $\{y|y \geq 0\}$ **4.** domain = ℝ, range = ℝ **5.** domain = $\{x \mid x \neq 0\}$

6. a. answers may vary **b.** Y=x should have a domain of all real numbers and $y = \sqrt{x}\sqrt{x}$ should have a domain of all non-negative real numbers. **c.** The table agrees with the statement made in part b. **d.** The graph agrees with the statement made in part b. With respect to viewing windows, answers may vary.

7. a. The domain should be $\{x| x \neq -3/2\}$. **b.** The domain is $\{x| x \neq -3/2\}$. **c.** The table will not support the declared domain unless you know to increment the table by ½. The graph supports the domain stated in part b. With respect to viewing windows, answers may vary.

8. B **9.** D **10.** A **11.** C

12. (.15, -5.205125), (-3/20, -41641/8000) **13.** (3.5, 1.414213562)

14. (-1/2, -8/27) **15.** $\sqrt{3X+5}$ is equivalent to $\sqrt{-19}$ when X=-8. The square root function is undefined for negative radicands.

16. a. $Y_1(2000)$ 5000 **b.** The $5,000 is positive, and therefore a profit.
c. $Y_1(1000)$ -5000 The negative indicates that the $5,000 would be a loss.
17. answers may vary

UNIT 17
TRANSLATING, STRETCHING, AND SHRINKING GRAPHS OF FUNCTIONS

*Prerequisite: Unit #16

We will begin by exploring the graphs of quadratic functions. A quadratic function is an equation in the form $y = a\,x^2 + bx + c$, where a, b, and c are real numbers, $a \neq 0$.

Note: The calculator's maximum or minimum feature was used to find the vertex of each parabola. This feature is found under the CALC menu on the TI-83/84 series and by pressing ***[MORE] [MATH]*** *and then the appropriate F key on the TI-86. Regardless of the type of calculator, the option functions in the same manner as the ROOT/ZERO feature. There may be roundoff error when using this feature.*

Function	Numerically	Graphically
$f(x) = x^2$ $f(x) = -x^2$	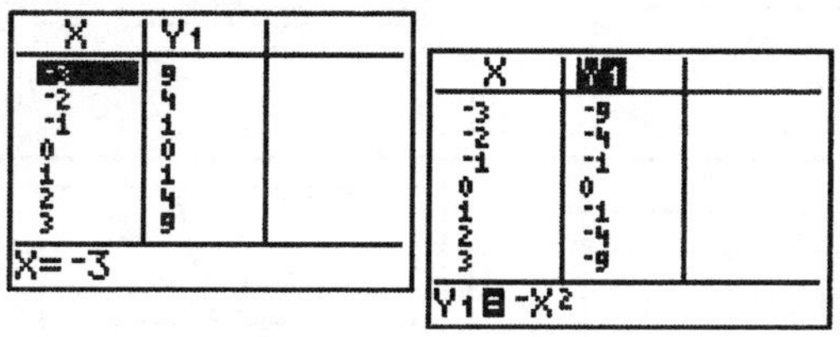	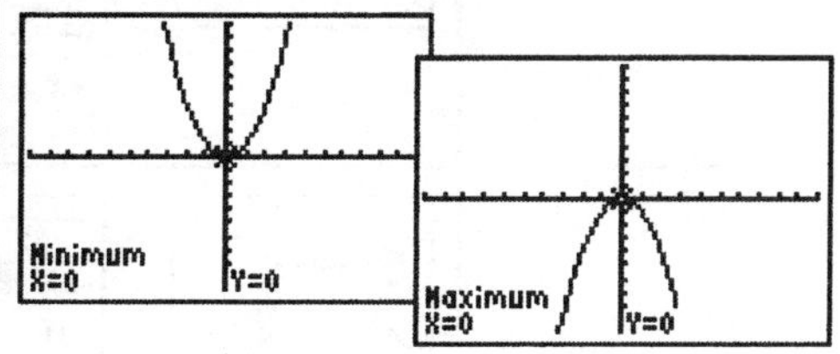
$f(x) = 3x^2$ $f(x) = -3x^2$	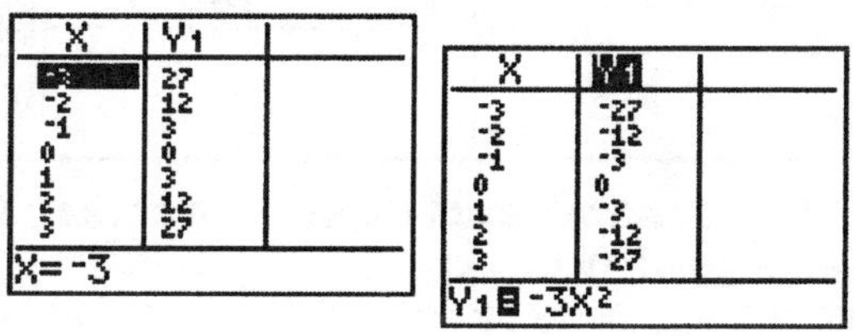	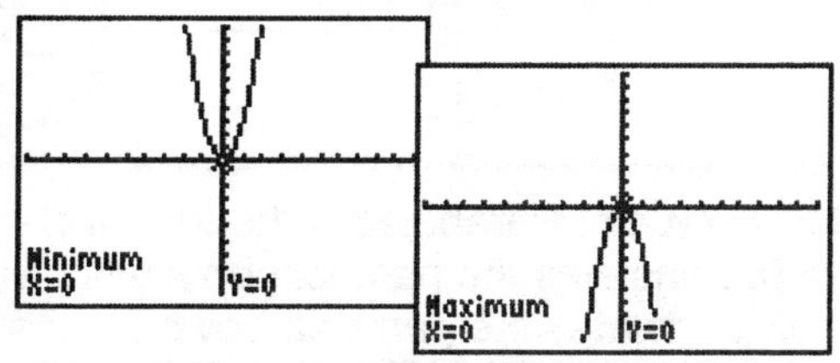
$f(x) = \frac{1}{4}x^2$ $f(x) = -\frac{1}{4}x^2$	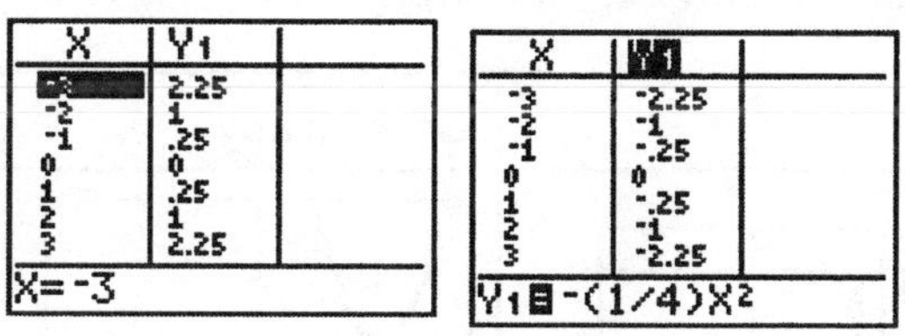	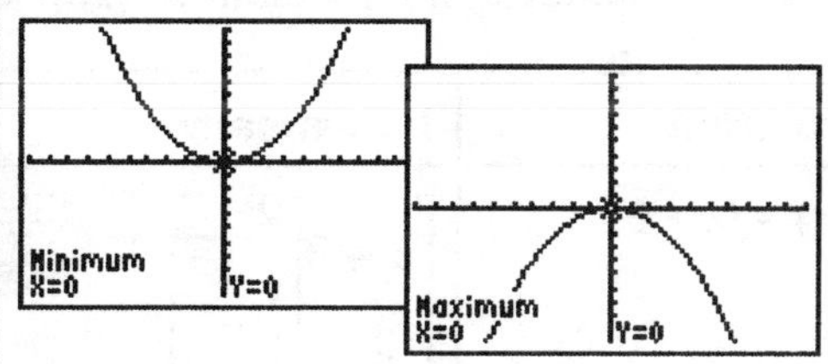

Observation: Parabolas of the form $f(x) = a\,x^2$ have a vertex of (0, 0) and a domain $\mathbb{R}$.
$a > 0$, parabola opens up and vertex is a minimum with a range of $\{y \mid y \geq 0\}$
$a < 0$, parabola opens down, vertex is a maximum with a range of $\{y \mid y \leq 0\}$
$0 < |a| < 1$ stretches parabola when compared to $f(x) = x^2$;
$|a| > 1$ compresses the parabola when compared to $f(x) = x^2$.

Consider the effects of a constant on graphs of the form $f(x) = x^2 + c$.

Function	Numerically	Graphically
$f(x) = x^2 + 3$	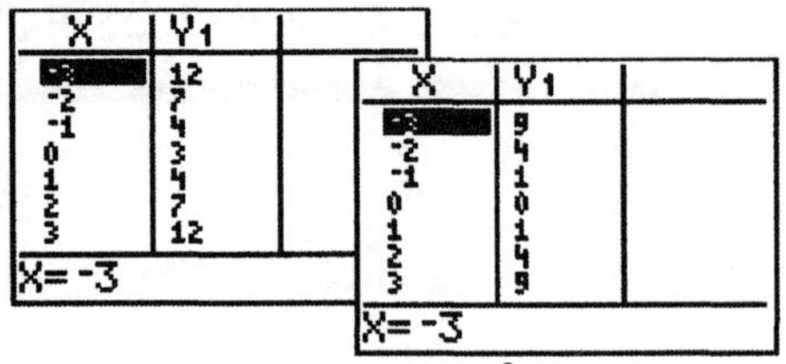 $f(x) = x^2$	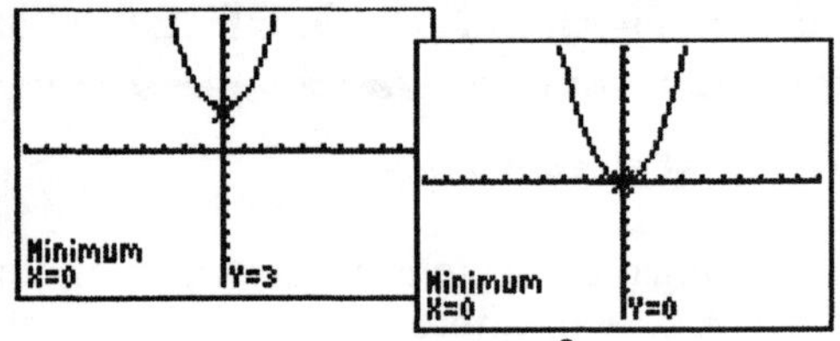 $f(x) = x^2$
$f(x) = x^2 + 1$	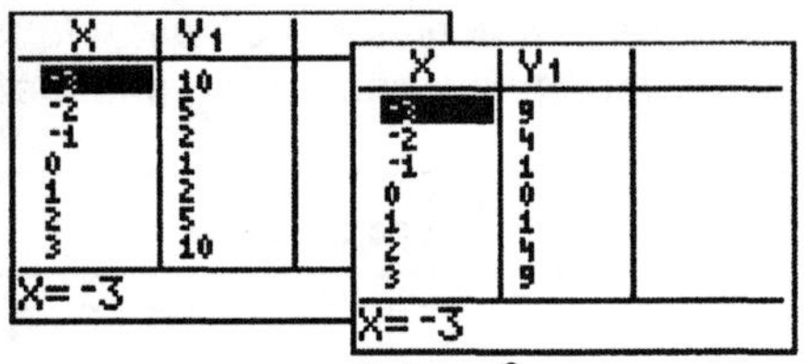 $f(x) = x^2$	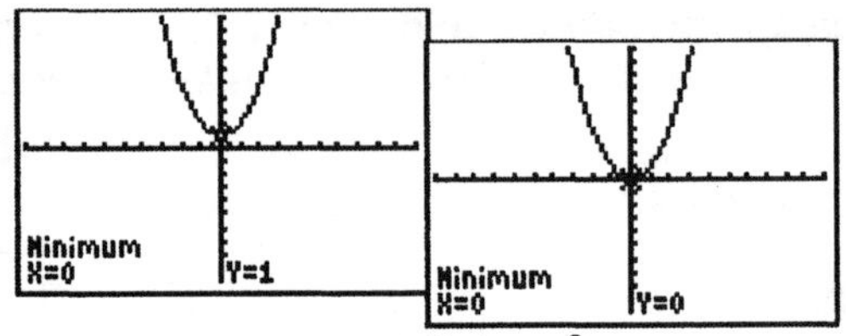 $f(x) = x^2$
$f(x) = x^2 - 2$	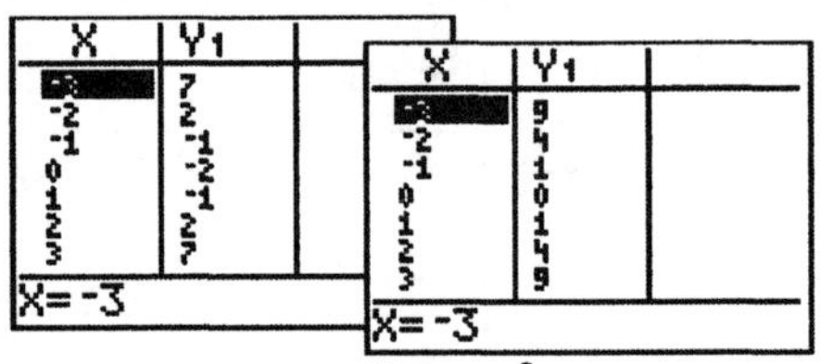 $f(x) = x^2$	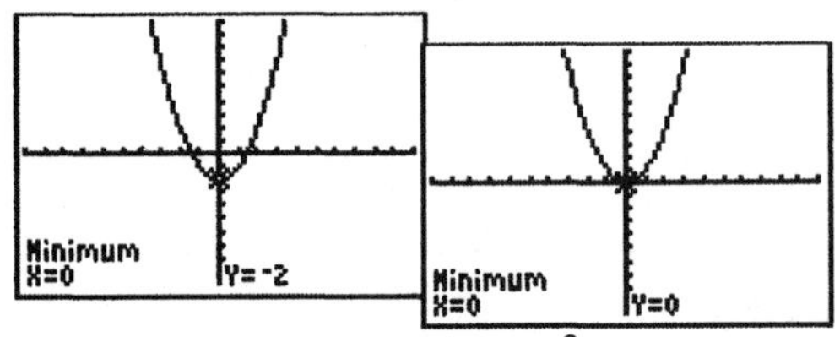 $f(x) = x^2$
Observation: Parabolas of the form $f(x) = x^2 + k$ have a vertex of (0, k) and a domain ℝ. k > 0, translates the parabola up k units with a range of $\{y \mid y \geq k\}$ k< 0, translates the parabola down k units with a range of $\{y \mid y \geq k\}$		

Consider the effects of a constant on graphs of the form $f(x) = (x - h)^2$.

Function	Numerically	Graphically
$f(x) = (x + 6)^2$	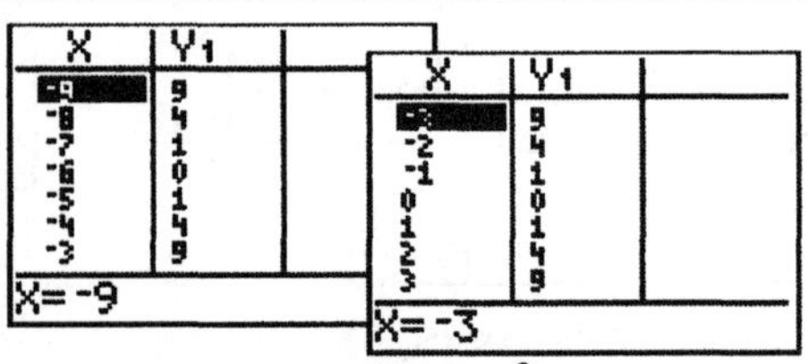 $f(x) = x^2$	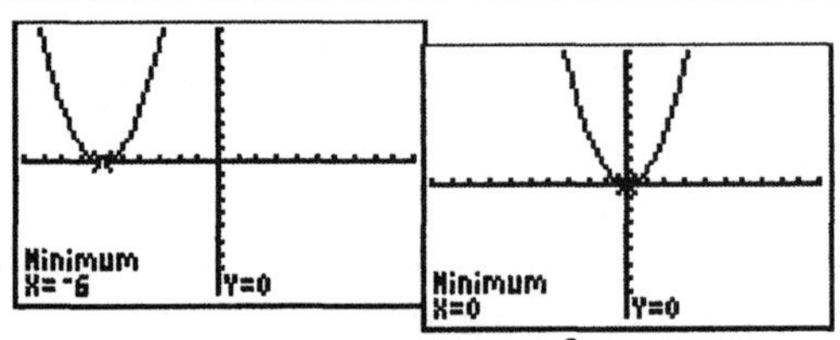 $f(x) = x^2$
$f(x) = (x + 1)^2$	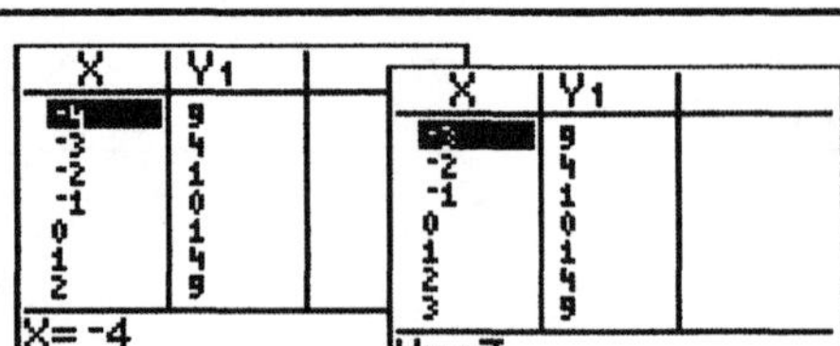 $f(x) = x^2$	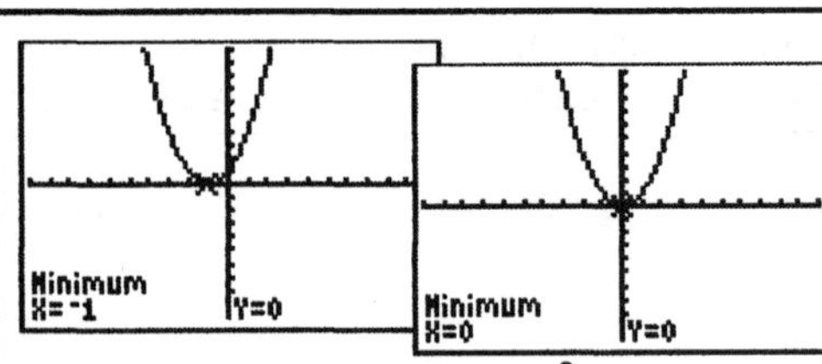 $f(x) = x^2$

$f(x) = (x - 3)^2$	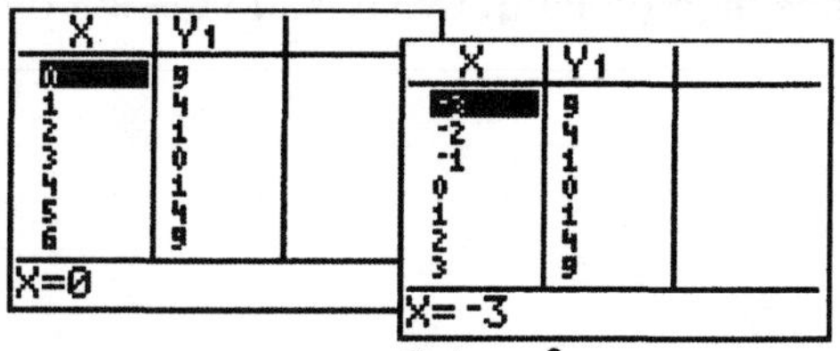 $f(x) = x^2$	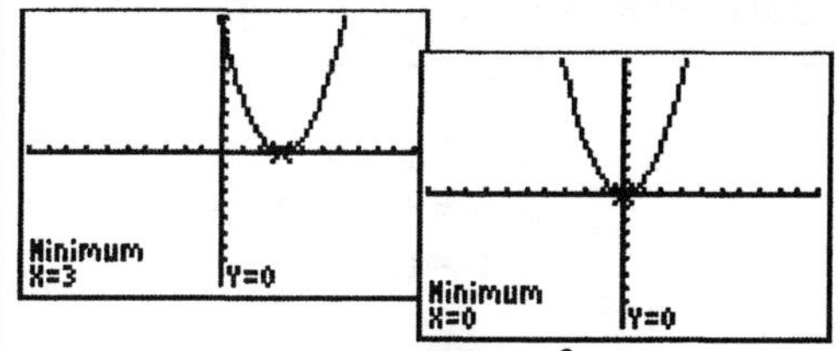 $f(x) = x^2$
Observation: Parabolas of the form $f(x) = (x - h)^2$ have a vertex of (h, 0) and a domain $\mathbb{R}$. h > 0, translates the parabola left h units with a range of $\{y \mid y \geq 0\}$ h< 0, translates the parabola right h units with a range of $\{y \mid y \geq 0\}$		

The constants in the general form of a parabolic equation, $f(x) = a(x - h)^2 + k$ translate (move the graph) in the rectangular coordinate system. The example below examines the effects of the constants a, h, and k.

Example: Given the function $f(x) = -2(x + 3)^2 - 4$. How does this graph compare to the graph of $f(x) = x^2$?

Solution: The parabola opens down because *a* is negative. The parabola will be compressed because $|a| > 1$. The graph shifts left three units and down 4 units. The vertex is (-3, -4).

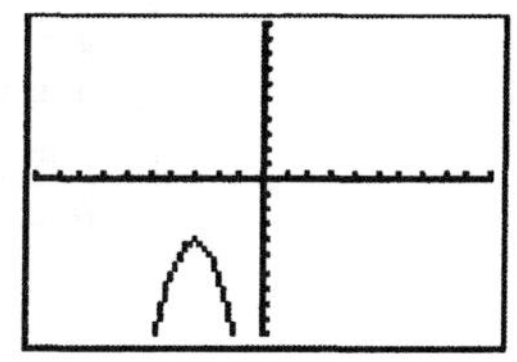

EXERCISE SET

Using what you have learned in this unit, match each graph below with one of the equations #1-4. ***Do not enter any expressions on the calculator.***

______1. $y = (x + 5)^2 - 4$

______2. $y = (x - 5)^2 + 4$

______3. $y = (x + 4)^2 + 5$

______4. $y = (x - 5)^2 - 4$

A.

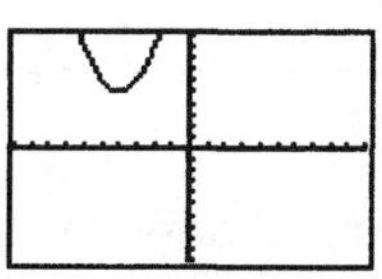

B.

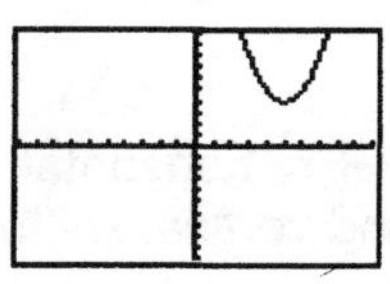

C.

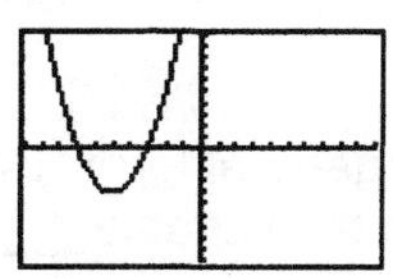

D.

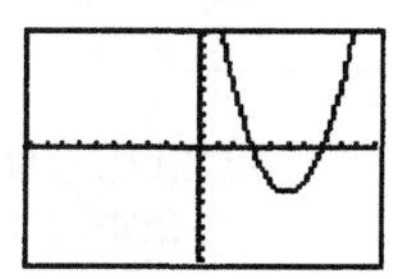

CONCLUSIONS: Answer the following questions without graphing the equation on the graphing calculator. You may then go back and check your answers with the calculator. Some questions may have more than one correct response.

_____5. Which of these graphs will have its vertex at the origin?

a. $y = (x - 5)^2$
b. $y = x^2$
c. $y = (4/5)y^2$
d. $y = 2x^2 + 7$
e. $y = 4(x + 3)^2 - 7$

____6. Which of these is the graph of $y = x^2$ translated (shifted) two units to the left of the y-axis?

a. $y = 2x^2$
b. $y = (x + 2)^2$
c. $y = x^2 - 2$
d. $y = (x + 2)^2 - 2$
e. $y = (x - 2)^2$

____7. Which of these is the graph of $y = x^2$ translated 2 units down from the x-axis?

a. $y = 2x^2$
b. $y = (x + 2)^2$
c. $y = x^2 - 2$
d. $y = (x + 2)^2 - 2$
e. $y = (x - 2)^2$

____8. Which of these graphs has a maximum point?

a. $y = 2x^2$
b. $y = -2(x - 2)^2$
c. $y = x^2 - 2$
d. $y = (x + 2)^2 - 2$
e. $y = 2 - x^2$

9. a. Using the standard viewing window, consider the following absolute value equations in the form $y = f(x) + c$. The graph of $f(x) = |x|$ has been graphed as the reference graph. Graph $y = |x| + 4$ and $y = |x| - 4$ on this same set of axes, labeling each graph.

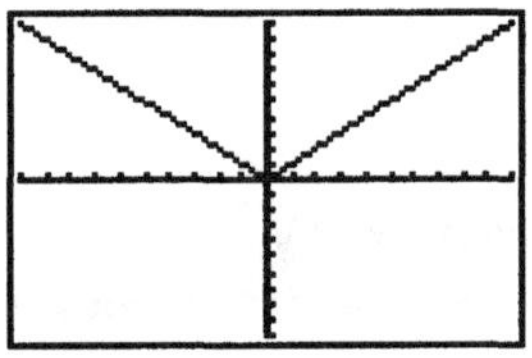

The range of $y = |x| + 4$ is ________________.
The range of $y = |x| - 4$ is ________________.
The domain for both graphs is $\mathbb{R}$.

b. The next group of graphs is in the form $y = f(x + b)$. The reference graph of $f(x) = |x|$ is already displayed. Graph $y = |x + 5|$ and $y = |x - 5|$ on this same set of axes, labeling each graph.
The domain for both graphs is and the range for both graphs is __________.

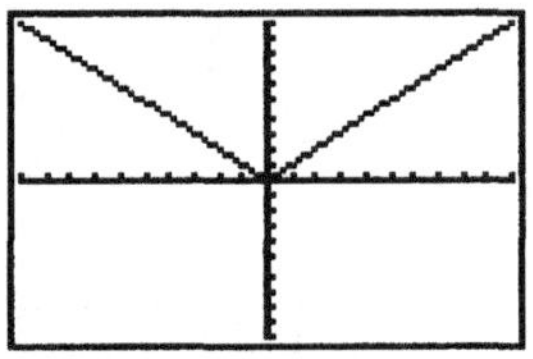

✍ c. CONCLUSION: In the general form $y = f(x + b) + c$, horizontal shifts result from changes in the variable _______ and vertical shifts result from changes in the variable ________ .

d. In the case of an absolute value function, does the horizontal shift ever affect the domain or range? __________

e. Does the vertical shift affect the domain or range? ________

f. Based on your conclusions, the graph of $y = |x + 32| - 42$ should translate vertically ______ units _______ (up/down) and horizontally _______ units ________ (left/right) when compared to the graph of $f(x) = |x|$.

This would locate the vertex of the absolute value function at the coordinates (______ , ______).

10. a. Using the same viewing window, consider the following square root functions in the form $y = f(x) + c$. The graph of $f(x) = \sqrt{x}$ has been graphed as the reference graph. Graph $y = \sqrt{x} + 4$ and $y = \sqrt{x} - 4$ on this same set of axes, labeling each graph.

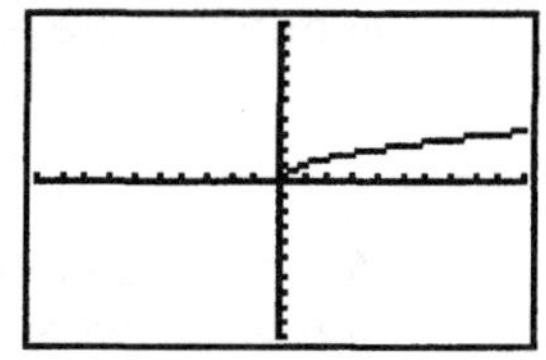

The range of $y = \sqrt{x} + 4$ is ____________________.

The range of $y = \sqrt{x} - 4$ is ____________________.

The domain of both graphs is ____________________.

b. The next group of graphs is in the form $y = f(x + b)$. The reference graph of $f(x) = \sqrt{x}$ is displayed.

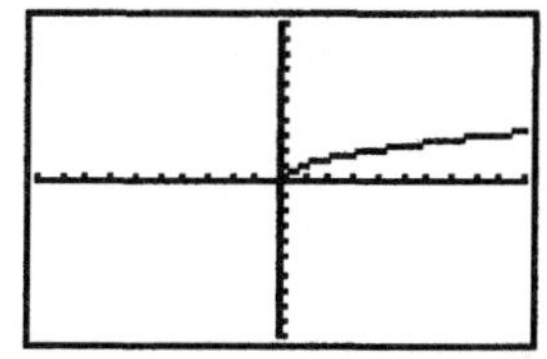

Graph $y = \sqrt{x+5}$ and $y = \sqrt{x-5}$ on this same set of axes, labeling each graph.

The domain of $y = \sqrt{x+5}$ is ____________________.

The domain of $y = \sqrt{x-5}$ is ____________________.

The range for both graphs is ____________________.

c. CONCLUSION: In the general form $y = f(x + b) + c$, horizontal shifts result from changes in the variable ________ and vertical shifts result from changes in the variable ________ .

d. In the case of a square root function, does the horizontal shift ever affect the domain or range? ____________

e. Does the vertical shift affect the domain or range? ___________

f. Based on your conclusions, the graph of $y = \sqrt{x+23} - 31$ should translate horizontally ______ units ________ (left/right) and vertically _______ units _______ (up/down) when compared to $f(x) = \sqrt{x}$.

This would locate the initial point of the square root curve at the coordinates (______ , ______).

11. Graph each of group of functions on the same set of axes. They are in the form $y = a \cdot f(x)$. The reference graph, listed first in the series, is graphed for you. Label each graph on the display.

$y = x^2$, $y = 4x^2$, $y = (½)x^2$

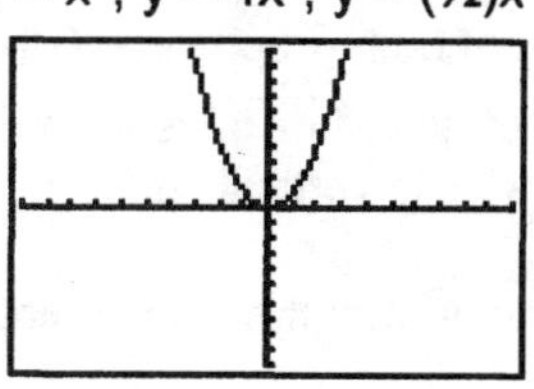

$y = |x|$, $y = 4|x|$, $y = (½)|x|$

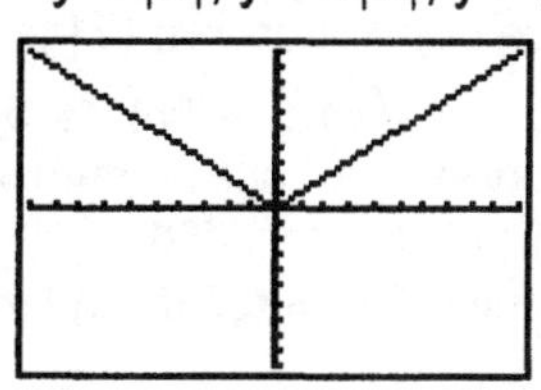

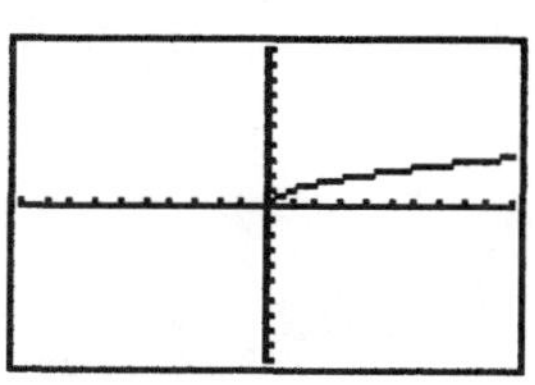

$y = \sqrt{x}$, $y = 4\sqrt{x}$, $y = (½)\sqrt{x}$

a. In general, if a > 0, what effect does **a** have on the graph of $y = a \cdot f(x)$?

b. What will happen to the graph of $y = a \cdot f(x)$ if a < 0? (If need be, go back to the specified problems and graph when a < 0 instead of a > 0.)

12. The graph of $f(x) = \sqrt{x}$ swings right and up, $f(x) = -\sqrt{x}$ swings right and down. What must be done to $f(x)$ so that the graph of the square root function swings left?

13. Write the equation of a square root function with a domain of $\{x \mid x \leq 1\}$ and a range of $\{y \mid y \geq -3\}$.

14. Write the equation of a square root function with a domain of $\{x \mid x \leq -1\}$ and a range of $\{y \mid y \leq 3\}$.

15. Write the equation of a square root function with a domain of $\{x \mid x \geq 4\}$ and a range of $\{y \mid y \leq 5\}$.

16. Write the equation of an absolute value function with a domain of $\mathbb{R}$ and a range of $\{y \mid y \leq 5\}$.

SOLUTIONS:

1. c **2.** b **3.** a **4.** d **5.** b,c **6.** b,d **7.** c,d **8.** b,e

9. a. range of $y = |x| + 4$: $\{y \mid y \geq 4\}$, range of $y = |x| - 4$: $\{y \mid y \geq -4\}$ **b.** In the next group of graphs the domain is $\mathbb{R}$ and the range is $\{y \mid y \geq 0\}$ for both graphs. **c.** Conclusion: "b" affects horizontal shift and "c" affects vertical shift. **d.** No **e.** The vertical shift affects only the range.
f. Thus $y = |x + 32| - 42$ translates vertically 42 units down and horizontally 32 units left with the vertex at (-32, - 42).

10. a. range of $y = \sqrt{x} + 4$: $\{y \mid y \geq 4\}$, range of $y = \sqrt{x} - 4$: $\{y \mid y \geq -4\}$. The domain of both graphs is $\{x \mid x \geq 0\}$.
b. In the next group of graphs the domain of $y = \sqrt{x+5}$ is $\{x \mid x \geq -5\}$ and the domain of $y = \sqrt{x-5}$ is $\{x \mid x \geq 5\}$. The range is $\{y \mid y \geq 0\}$ for both graphs. **c.** Conclusion: "b" affects horizontal shift and "c" affects vertical shift.
d. Only the domain is affected. **e.** Only the range is affected.
f. Thus $y = \sqrt{x+23} - 31$ translates horizontally 23 units left and vertically 31 units down with the initial point of the curve at (- 23,- 31).

11. a. "a" affects the "width" of the graph: narrows it. **b.** If a < 0 the graph will be symmetric across the x-axis to the graph whose "a" is positive.

12. The opposite of the radicand must be graphed. **13.** $f(x) = \sqrt{1-x} - 3$

14. $f(x) = -\sqrt{-1-x} + 3$ **15.** $f(x) = -\sqrt{x-4} + 5$ **16.** $f(x) = -|x| + 5$

UNIT 18
SYMMETRY OF FUNCTIONS

*Prerequisite: Unit #17

Finding the x and *y*-intercepts is helpful in adjusting the window values on the calculator so that a complete graph is displayed. Also helpful in sketching (and determining a good viewing window) is the concept of symmetry. It is suggested that you set your window values to ZDecimal x 2, as specified beneath the first set of displayed graphs below.

Symmetry About the *y-axis*

Algebraically	**Geometrical Interpretation**	**Graphically**	**Numerical**
Is the graph of $y = x^2 - 4$ symmetric with respect to the *y*-axis? Is $f(x) = f(-x)$?	If the graph is folded along the y-axis, does the left side match the right side exactly? To confirm this, the points of the curve will have ordered pairs of the form (x, y) and $(-x, y)$. (see the graphical display and table display)	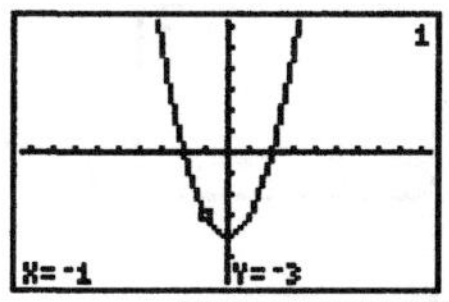 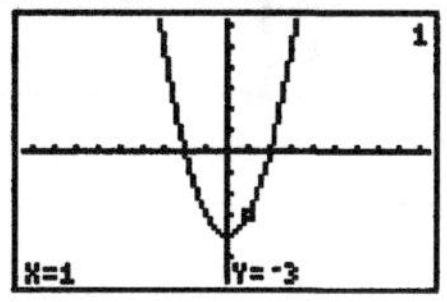 Window values: [-9.4,9.4] by [-6.2, 6.2] TI-86: [-12.6, 12.6] by [-6.2, 6.2]	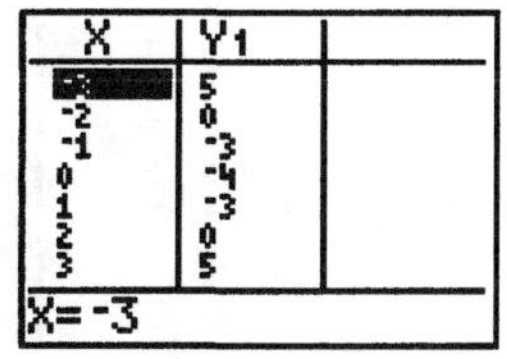 The symmetry becomes apparent numerically. This confirms, but *does not prove*, that the graph of $y = x^2 - 4$ is symmetric with respect to the y-axis

Symmetry About the Origin

Algebraically	**Geometrical Interpretation**	**Graphically**	**Numerically**
Is the graph of $y = \frac{3}{x}$ symmetric with respect to the origin? Is $f(-x) = -f(x)$?	Imagine rotating the graph 180° about the origin. The "top" part of the graph would line up with the "bottom" part if there is symmetry about the origin. Points on the curve will have ordered pairs of the form (x, y) and $(-x, -y)$.	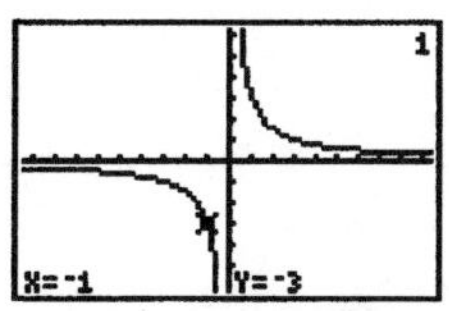 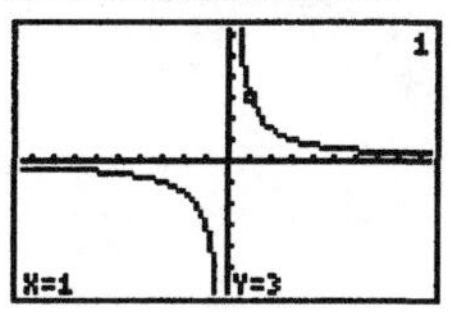	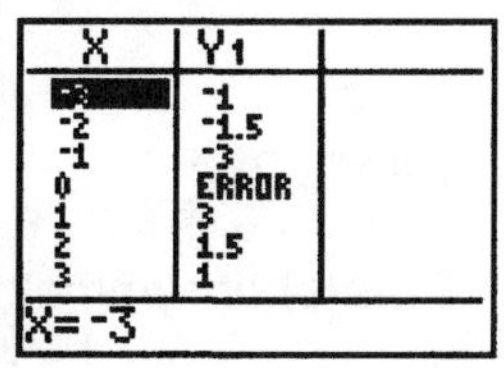 Because coordinates of the form (x, y) and $(-x, -y)$ are on the curve, the table confirms that the graph is symmetric with respect to the origin.

EXERCISE SET

1. Each pictured graph below is symmetric with respect to the *y*-axis. Using the information displayed at the bottom of the screen, state the coordinates of another point which also lies on the graph.

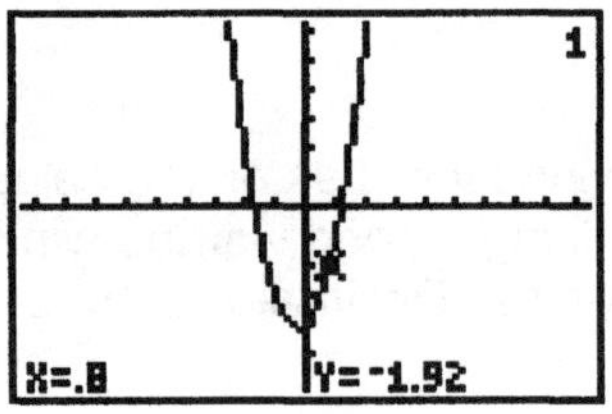

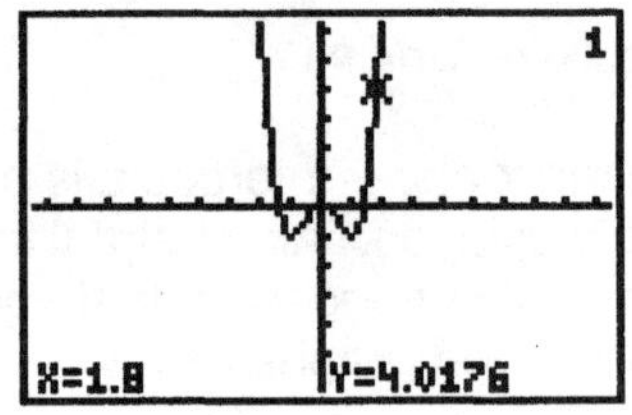

Coordinates:______________ Coordinates:______________

2. Each pictured graph below is symmetric with respect to the origin. Using the information displayed at the bottom of the screen, state the coordinates of another point which also lies on the graph.

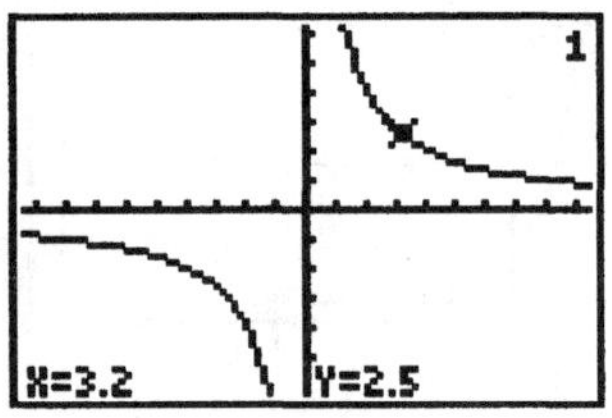

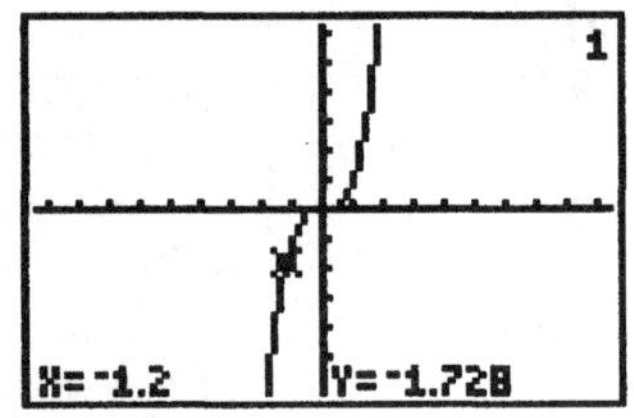

Coordinates:______________ Coordinates:______________

Directions: Sketch the graph of each of the following polynomial functions on the grid provided. Use a ZDecimal x 2 screen ([-9.4,9.4] by [-6.2, 6.2] for theTI-83/84 series and [-12.6, 12.6] by [-6.2, 6.2] for the TI-86). State whether the function is symmetric with respect to the *y*-axis, and/or with respect to the origin, or has no symmetry. Justify the symmetry (or lack of) by stating the coordinates of three <u>pairs</u> of points whose coordinates confirm the *y*-axis symmetry or symmetry about the origin.

3. $y = |x|$

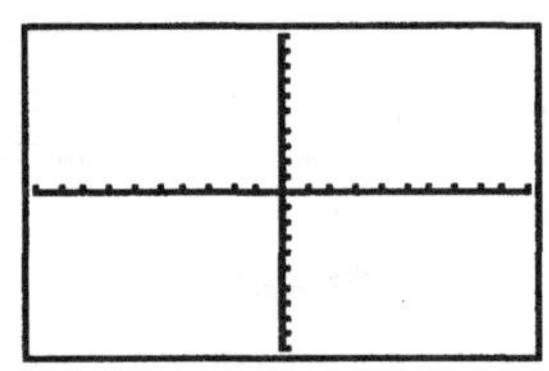

Type(s) of symmetry:

Justification:

4. $y = x^4 - 2x^2$

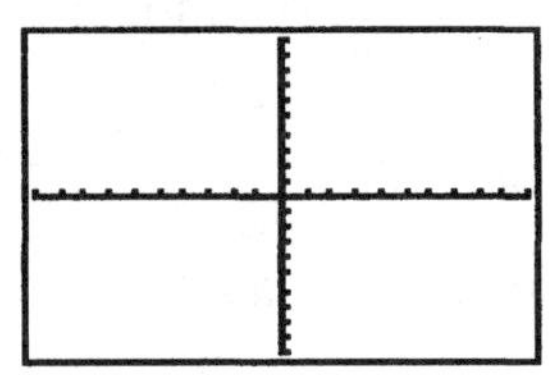

Type(s) of symmetry:

Justification:

5. $y = x^4 + x^3 - 2x^2$

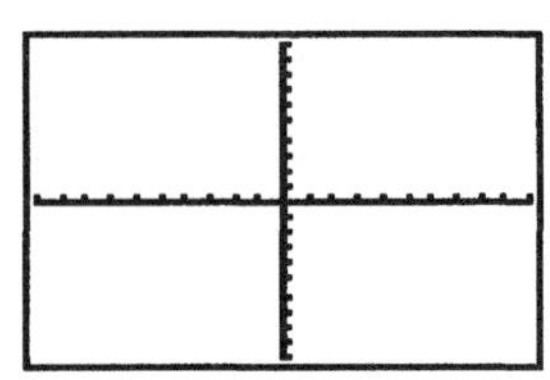

Type(s) of symmetry:

Justification:

6. $xy = 8$

Type(s) of symmetry:

Justification:

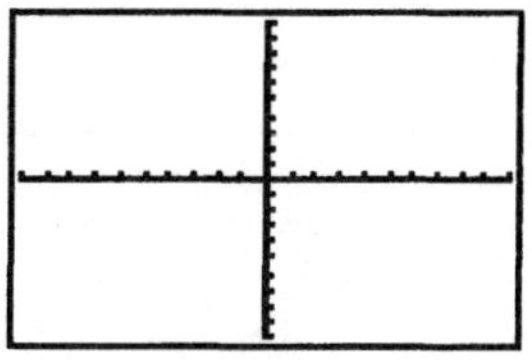

7. $y = x^3 - x$

Type(s) of symmetry:

Justification:

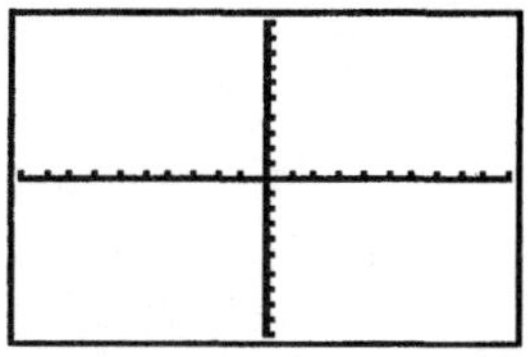

8. $y = x^3 + 3$

Type(s) of symmetry:

Justification:

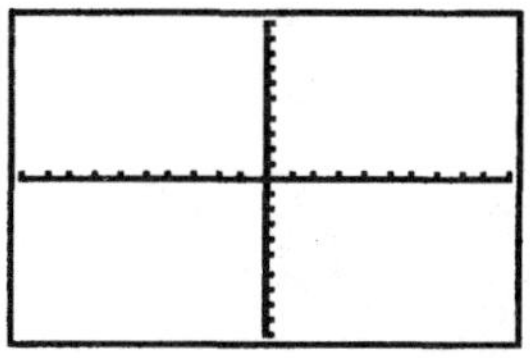

✍9. A function that is symmetric about the *y*-axis is called an <u>even</u> function whereas one whose graph is symmetric about the origin is called an <u>odd</u> function. Use the calculator to explore polynomial functions. Consider graphs of polynomial functions whose powers of the variable *x* are (a) even, (b) odd, (c) a combination of even and odd. What conclusions (if any) can be made about these functions and symmetry?

✍10. This unit has only examined symmetry about the *y*-axis and the origin. Another type of symmetry is symmetry about the *x*-axis. The graphs of these equations are <u>NOT</u> functions. Explore symmetry about the *x*-axis by graphing $y^2 = x$. (Hint: You must graph TWO curves that together **represent** the graph of $y^2 = x$.) In conclusion, specify the form taken by the ordered pairs that satisfy this relation.

Solutions:

1. (-.8,-1.92) and (-1.8,4.0176)

2. (-3.2,-2.5) and (1.2,1.728)

Type of Symmetry	Justification (points may vary)
3. Y-axis	(-3,3) (-2,2) (-1,1) (3,3) (2,2) (1,1)
4. Y-axis	(-3,63) (-2,8) (-1,-1) (3,63) (2,8) (1,-1)
5. None	(-3,36) (-2,0) (-1,-2) (3,90) (2,16) (1,0)
6. Origin	(-3,-2.667) (-2,-4) (-1,-8) (3, 2.667) (2, 4) (1, 8)
7. Origin	(-3,-24) (-2,-6) (-1,0) (3, 24) (2, 6) (1,0)
8. None	(-3,-24) (-2,-5) (-1,2) (3, 30) (2,11) (1,4)

9. Answers may vary. **10.** Answers may vary.

UNIT 19
PIECEWISE FUNCTIONS

*Prerequisite: Unit #14

This unit examines piecewise functions; functions in which $f(x)$ varies for different intervals of the domain. In these functions, the domain is segmented into a finite number of pieces. The **TEST** menu will be used to graph the different pieces of the function. Because we do not want the pieces connected, set the calculator mode to dot at this time.

You have the option of changing to **DOT** on the **MODE** screen or simply setting the graphing style icon on the **Y=** menu to DOT for each equation graphed. In either case, the graphing style icon should reflect the dot graphing format and should be verified each time a new graph is displayed.

Test Menu

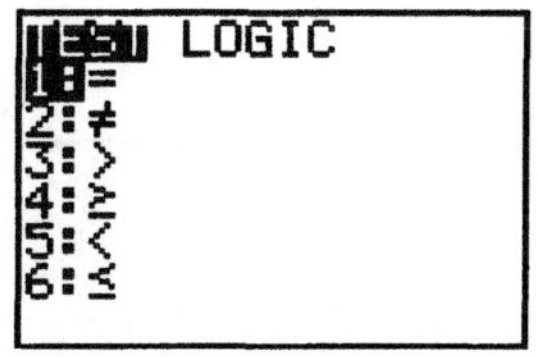

The TEST menu is located above the MATH key (TI-86: above the digit 2). When an inequality symbol from the TEST menu is designated in the function, the calculator evaluates the function for all real numbers in the domain. For each value tested that ***is*** in the designated interval, the calculator returns a **1** and for each value tested that ***is not*** in the designated interval the calculator returns a **0**. Thus, the **1** turns the point **on** and allows it to be displayed, whereas, the **0** turns the point **off** so it is not displayed as part of the function. The following examples demonstrate the calculator's response.

Example 1	**Numerically**	**Graphically**
Graph the piecewise function $f(x) = x + 2$ when $x \le 5$ on the grid provided.	Because the domain is $x \le 5$, the graph of the equation will **begin** at $x = 5$ and pass through all points whose x value is less than or equal to five. Plot the ordered pairs and draw the **ray** which represents the graph of $f(x) = x + 2$ when $x \le 5$. We used (5,7),(4,6) and (3,5) to determine the line.	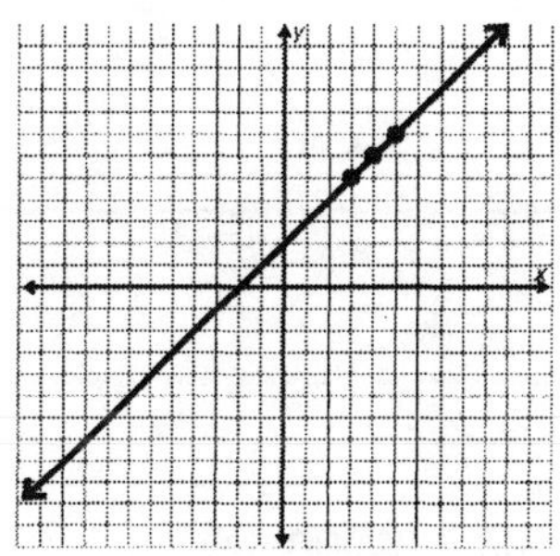

Example 2	**Graphically**	
How does the calculator plot ordered pairs for the function $f(x) = x + 2$ when $x \le 5$?	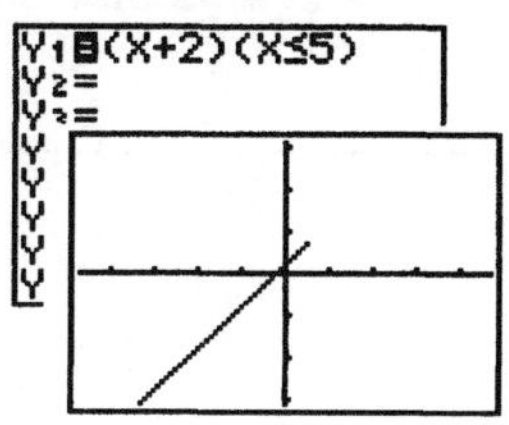	Enter the function at the Y1 prompt, designating the restriction on the domain, $x \le 5$, as indicated. To plot points, the calculator substitutes values for x and evaluates the function to determine the corresponding y values. When an x value is substituted into the **TEST** expression ($x \le 5$), the calculator **tests** to see if that value is **less than** or **equal** to 5. If it is less than or equal to 5, the calculator turns the point *on* and plots it. If the value is **greater than** 5, the calculator turns the point *off*, preventing it from being displayed as part of the function. The notation for *on* and *off* are **1** and **0**, respectively. With the viewing window in ZInteger, press **[GRAPH]** to display the screen.

Numerical Demonstration of Calculator's Response to Various Values of x					Notes
x	(x + 2)(x ≤ 5)	Point on/off?	f(x)	ordered pair	Observe that the last three *x*-values (6,7,8) returned a zero and were turned off since they were not part of the domain, *x* ≤ 5. When all points are graphed, where *x* ≤ 5 the desired graph is displayed. Points that are not part of the function have a *y*-value of 0 and are plotted on the *x*-axis where their display is not apparent.
3	(3 + 2)(1)	ON	5	(3,5)	
4	(4 + 2)(1)	ON	6	(4,6)	
5	(5 + 2)(1)	ON	7	(5,7)	
6	(6 + 2)(0)	**OFF**	0	(6,0)	
7	(7 + 2)(0)	**OFF**	0	(7,0)	
8	(8 + 2)(0)	**OFF**	0	(8,0)	

Example 3	**Graphically**	
Graph the function *f*(*x*) = *x* + 2 when *x* > 5	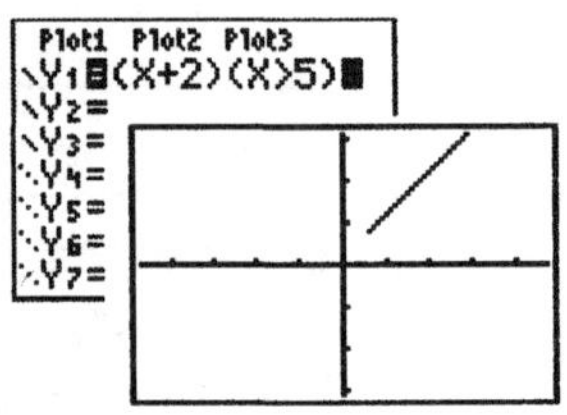	The display is the section of *f*(*x*) = *x* + 2 when *x* > 5.

Example 4	**Graphically**	
Graph the function *f*(*x*) = *x* + 2 when -5 < *x* < 5	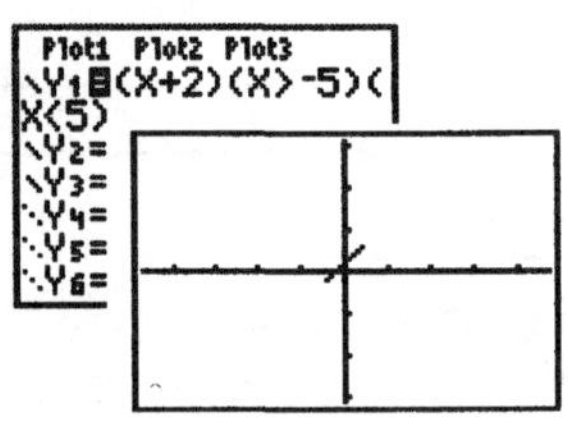	The display you see is the section of *f*(*x*) = *x* + 2 when - 5 < *x* < 5. This would be read: *x* is greater than negative five (*x* > -5) **AND** *x* is less than five (*x* < 5).

EXERCISE SET

Directions: Without graphing, match each of the piecewise functions defined below with their graph. All pictured graphs are in the ZStandard viewing window.

______1. $f(x) = x^2 + 1$ when $x \geq 0$

______2. $f(x) = \begin{cases} \sqrt{x} + 2 & \text{when } x \geq 0 \\ 2x & \text{when } x < 0 \end{cases}$

______3. $f(x) = \begin{cases} 3 & \text{when } x \geq 0 \\ -3 & \text{when } x < 0 \end{cases}$

______4. $f(x) = \begin{cases} x + 4 & \text{when } x \geq 2 \\ -2x - 5 & \text{when } x \leq 0 \end{cases}$

a.

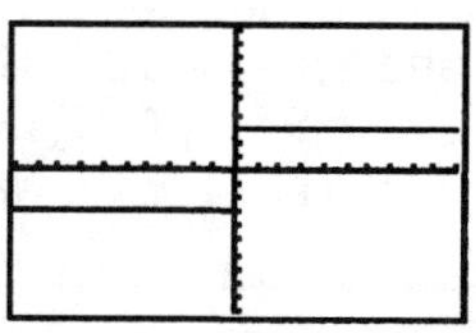

b.

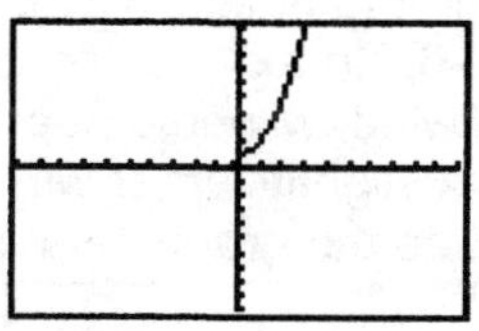

c.

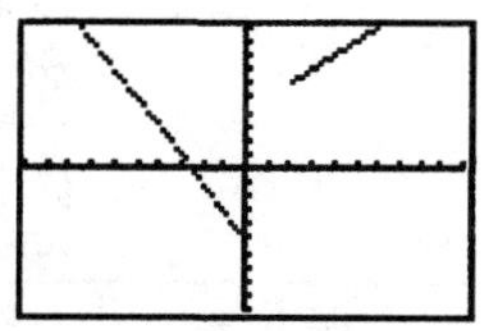

d.

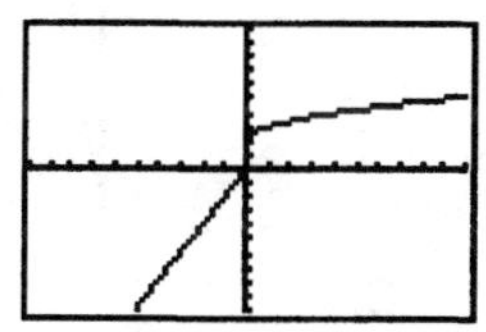

Directions: Graph each piecewise function in the ZStandard viewing WINDOW.

5. $f(x) = -2x^2 + 5$ when $x < 1$

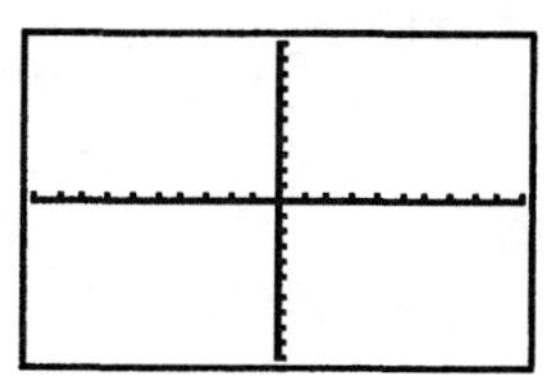

6. $f(x) = |x + 3|$ when $-6 < x < 0$

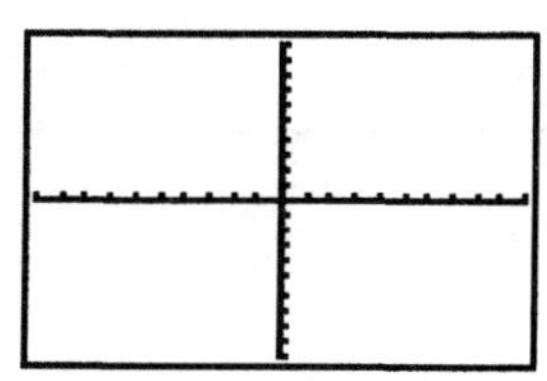

Setting Windows

When setting window values for piecewise functions, the Ymin and Ymax values should be determined by the methods discussed in the unit titled *Graphing Basics*. The *x*-values will, however, be determined by the restrictions placed on the domain in the problem. In the first example, $f(x) = x + 2$ when $x \le 5$. Thus the Xmax would be at least 5 and the minimum would need to be a value which would yield a complete or satisfactory graph (as explained in *Graphing Basics*). Since $f(x) = x + 2$ is a linear function, it would be acceptable to set the Xmin to -5. The next two examples consider other possibilities.

Example 5	**Graphically**
Determine the viewing window for $f(x) = (x + 6)^2 + 4$ when $x < -3$.	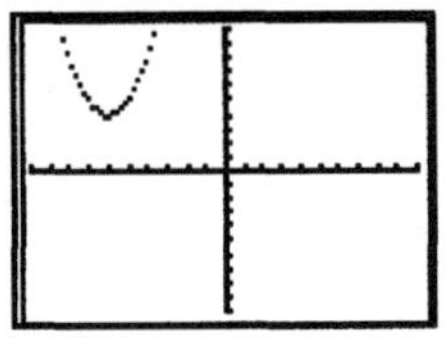 Graph the function in the ZStandard window as displayed. At the **Y** prompt enter **((x + 6)^2 + 4)(x<-3).** Compare your graph to the one pictured. How can you be confident that you are seeing a satisfactory graph of the **piecewise** function? To assure a satisfactory graph of piecewise functions you must consider the effect of the domain on the function, as well as those points which display the graph's interesting features. 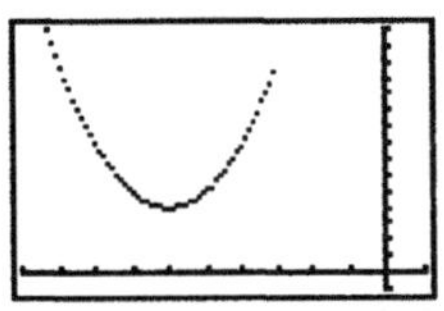 The Xmax should be at least -3 since the domain is restricted to all $x < -3$. Since this a parabolic function, with vertex at (-6,4), the Xmin needs to be less than -6. Remember, a satisfactory graph displays all the interesting features of the graph. The two interesting features of this graph are its vertex and the point at which the curve terminates. Thus, be sure to include the *y*- values that correspond to the maximum/minimum values in the domain. Since x < -3 the graph would need to include the point (-3,13). It can be concluded that: Xmin < -6, Xmax > -3, Ymin < 4 and Ymax > 13. The viewing window would need to be at least [-6,-3] by [4,13]. A slightly larger window: [-10,1] by [-1,15] was used. See the graph displayed.

Example 6	**Graphically**
Consider the graph of the piecewise function: $f(x)=\begin{cases} x^2+4x \text{ when } -4 \le x \prec 0 \\ 5 - 2x \quad \text{when } 0 \le x \prec 3 \end{cases}$	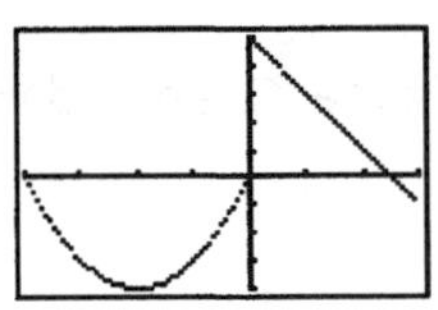 The function should be read as $f(x) = x^2 + 4x$ when $-4 \le x < 0$ **and** $f(x) = 5 - 2x$ when $0 \le x < 3$. This will be entered at the Y1 prompt as $(x^2+ 4x)(x\ge-4)(x< 0)+(5 -2x)(x \ge0)(x< 3)$. Press **[GRAPH]** to display the function. The screen displays the function when graphed in the smallest applicable viewing window,[-4,3] by [-4,5]. A satisfactory graph is displayed if the selected window displays the two pieces of the function shown. See the solutions key at the end of the unit for specific directions on selecting an appropriate viewing window.

EXERCISE SET CONTINUED

Directions: For each exercise below, complete the following

a. Determine the Xmin, Xmax, Ymin and Ymax values of the function.
b. Graph the function using the TEST menu.
c. State the final viewing WINDOW in the format [Xmin,Xmax] by [Ymin,Ymax].

7. $f(x) = \begin{cases} -2x^2 + 15 & \text{when } x \geq 0 \\ |x + 3| & \text{when } -12 < x < 0 \end{cases}$

Xmin = ______ Xmax = ______

Ymin = ______ Ymax = ______

WINDOW: [___, ___] by [___, ___]

8. $f(x) = \begin{cases} x^2 + 4x & \text{when } -4 \leq x < 0 \\ 5 - 2x & \text{when } 0 \leq x < 3 \\ x - 2 & \text{when } 3 \leq x < 6 \end{cases}$

Xmin = ______ Xmax = ______

Ymin = ______ Ymax = ______

WINDOW: [___, ___] by [___, ___]

9. The definition of absolute value specifies that $|x| = \begin{cases} x & \text{when } x \geq 0 \\ -x & \text{when } x \leq 0 \end{cases}$.

Using this definition, write the function $f(x) = |x + 4|$ as a piecewise function.

$f(x) = \begin{cases} \_\_\_\_\_\_\_\_\_\_\_\_\_\_\_\_ \\ \_\_\_\_\_\_\_\_\_\_\_\_\_\_\_\_ \end{cases}$

Establish an appropriate viewing window and sketch the graph display.

WINDOW: [___, ___] by [___, ___]

Your piecewise function should be entered after the Y1 prompt. To check the accuracy of your piecewise functon, enter |x + 4| after the Y2 prompt. Display both Y1 and Y2. You can verify they are identical by comparing the *y*-values in the TABLE feature.

10. Explore graphing greatest integer functions using the graphing calculator.

Solutions: **1.** b **2.** d **3.** a **4.** c **5.** At the Y1 prompt, enter $(-2x^2 + 5)(x<1)$

6. At the Y1 prompt, enter $(\text{abs}(x + 3))(x>-6)(x<0)$.

7. At the Y1 prompt, enter $(-2x^2 +15)(x\geq 0) + (\text{abs}(x + 3))(x>-12)(x< 0)$, Xmin = -12 and Xmax = ∞, since $x \geq 0$, Ymin = -∞ and Ymax = 15. Since (0,15) is the vertex of the quadratic, the window should be at least [-12,5] by [-10,16].

8. At the Y1 prompt, enter $(x^2 + 4x)(x\geq -4)(x< 0) + (5 - 2x)(x \geq 0)(x< 3) + (x - 2)(x \geq 3)(x < 6)$, Xmin = -4, Xmax =6, Ymin= -∞, Ymax = 5; the window should be at least [-4,6] by [-4,5].

9. $f(x) = x+ 4$ when $x \geq - 4$ and $f(x) = -x - 4$ when $x <- 4$. **10.** Answers may vary.

Example 6: Begin by determining the Xmax and Xmin for your viewing window as discussed in the previous example. Compare the domains: $-4 \leq x < 0$ and $0 \leq x < 3$. Taking the union of these two domains gives $-4 \leq x < 3$ and thus the Xmin and Xmax values are determined. (NOTE: Xmin can be less than -4 and Xmax can be greater than 3 if you wish.) We need to take into consideration the type of functions we are graphing as we consider the Ymax and Ymin . The function $f(x)=5 - 2x$ is linear and its only "interesting features" are its starting and stopping points. The starting point would be f(0), since $0 \leq x < 3$ is the domain. Thus our window needs to include (0,5). The stopping point would be f(3) which indicates our window should include (3,-1). Based on $f(x) = 5 - 2x$: Ymax ≥ 5, Ymin ≤ -1. DO NOT forget to take into consideration the graph of $f(x) =x^2 + 4x$. This is a parabolic function. Its "interesting features" will be its starting and stopping points, as well as its vertex. Since the domain is $-4 \leq x < 0$, f(-4) indicates the starting point is (-4,0) and f(0) indicates the stopping point is (0,0). The vertex is (-2,-4). Based on $f(x) = x^2 + 4x$: Ymax > 0, Ymin < -4 CONCLUSION: Ymax ≥5 and Ymin < -4. A suggested viewing window would be [-4,3] by [-4,5]. (Your viewing window may be larger than the suggested window but not smaller.)

UNIT 20
EXPONENTIAL AND LOGARITHMIC FUNCTIONS

*Prerequisite: Unit #16

This unit will examine the graphs of exponential functions (to include base *e*) and their inverses, logarithmic functions, and both exponential and logarithmic equations. All graphs - unless specified otherwise - should be displayed on the ZDecimal x 2 viewing window: i.e. [-9.4, 9.4] by [-6.2, 6.2], with both scales set to 1 for TI-83/84 series (TI-86 users: [-12.6,12.6] by [-6.2,6.2]).

Exponential Functions

Algebraically	Verbally	Numerically	Graphically
$f(x) = b$ where $b \neq 1$ and $b > 0$.	An exponential function has a positive constant base (other than one) and a variable exponent.	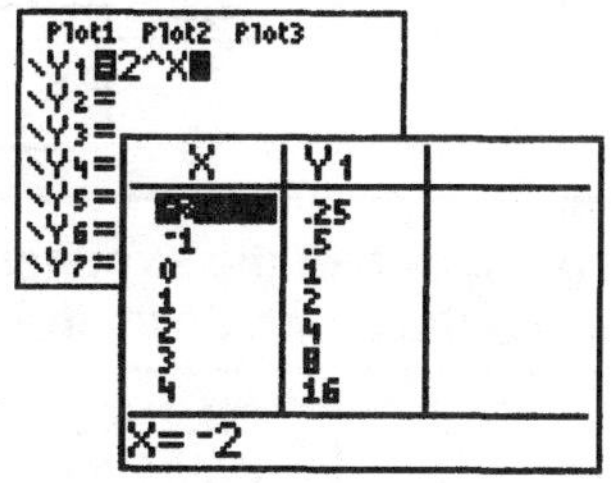	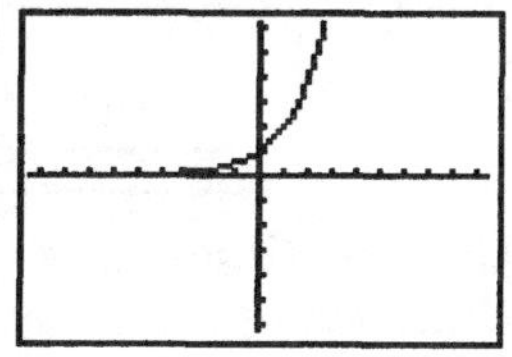

The following table examines different exponential functions.

Algebraically	Numerically	Graphically	Observations
$f(x) = 3^x$	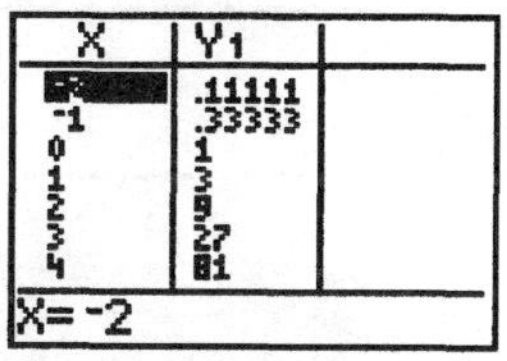	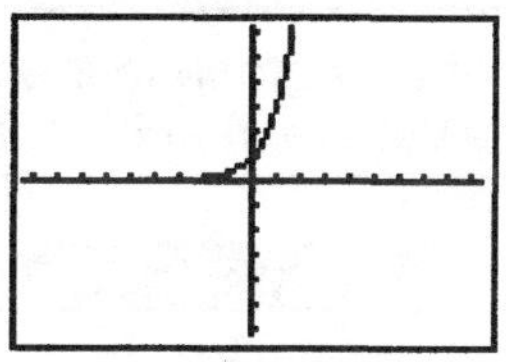	For $f(x) = b$, $b > 1$ The domain is $(-\infty, \infty)$; the range is $(0, \infty)$. There are no *x*-intercepts; the *y*-intercept is 1. The *x*-axis is a horizontal asymptote. The graph contains the points (0, 1) and (1, *b*). The function is increasing.
$f(x) = 8^x$	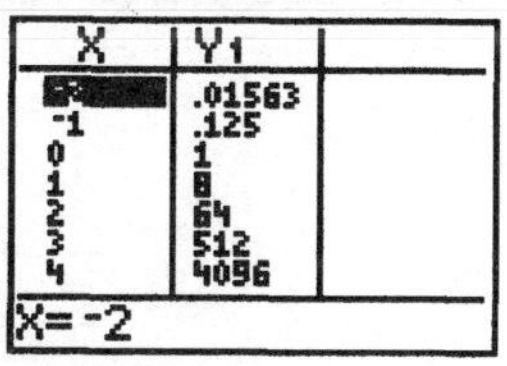	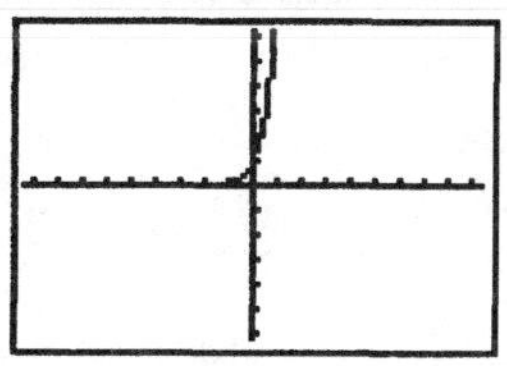	
$f(x) = \left(\frac{1}{2}\right)^x$	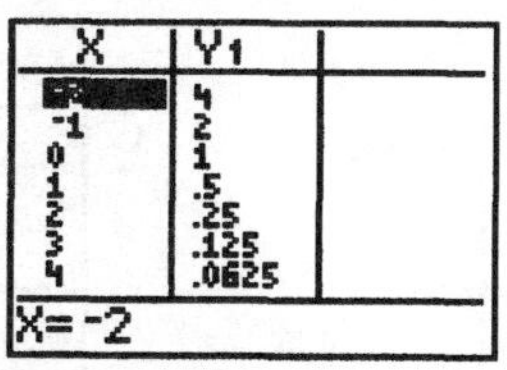	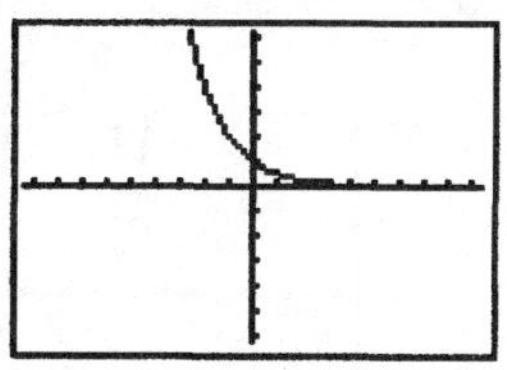	For $f(x) = b$, $0 < b < 1$ The domain is $(-\infty, \infty)$; the range is $(0, \infty)$. There are no *x*-intercepts; the *y*-intercept is 1. The *x*-axis is a horizontal asymptote. The graph contains the points (0, 1) and (1, *b*). The function is decreasing.
$f(x) = \left(\frac{1}{3}\right)^x$	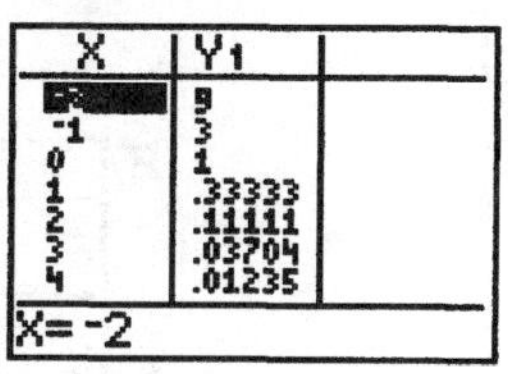	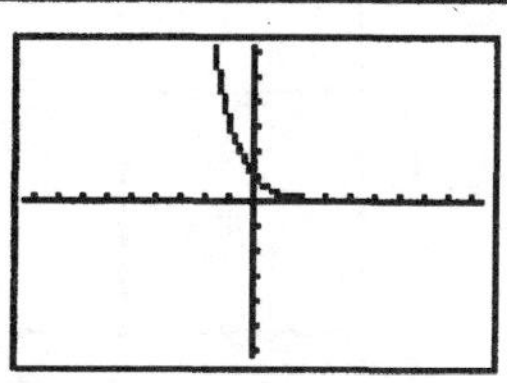	

$f(x) = \left(\frac{1}{8}\right)^x$

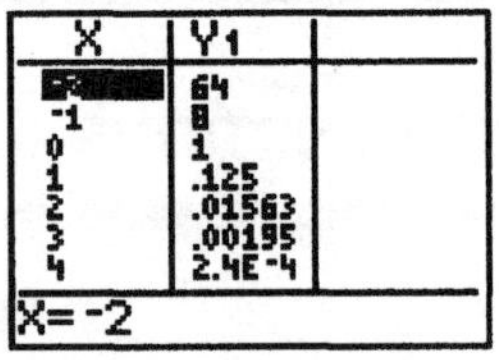

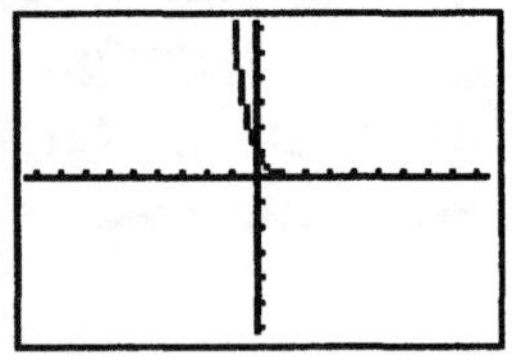

Transformation of Exponential Functions

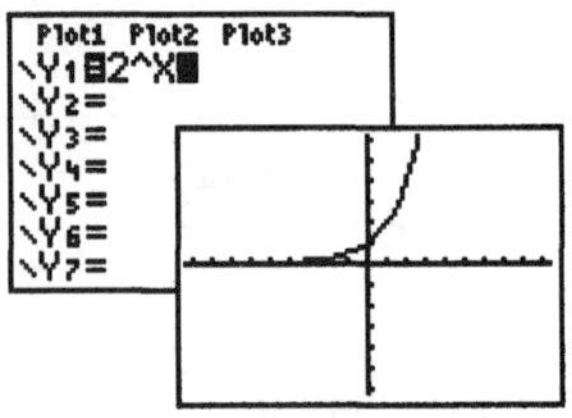

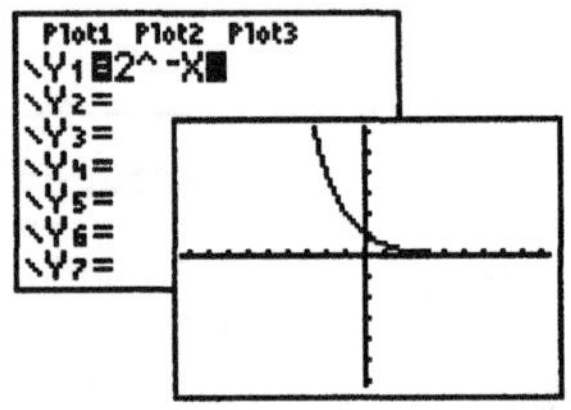

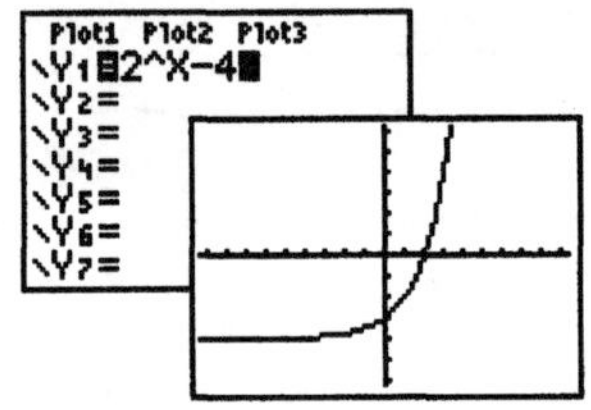

Observations: Changing the sign of the exponent reflects the graph about the *y*-axis. Subtracting a constant shifts the graph down the specified number of units.

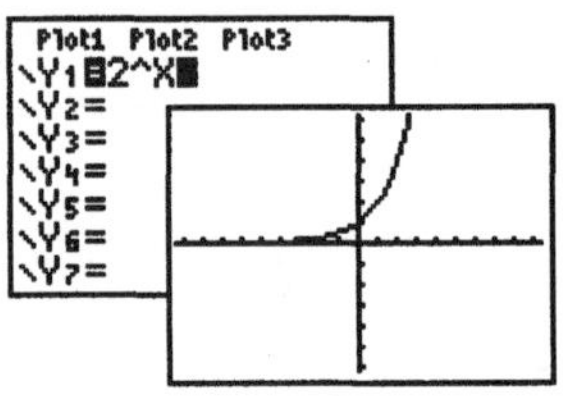

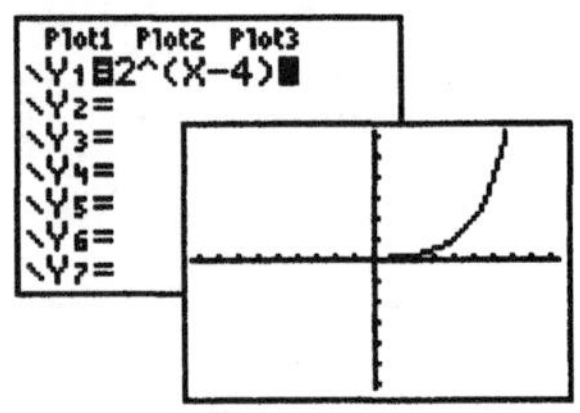

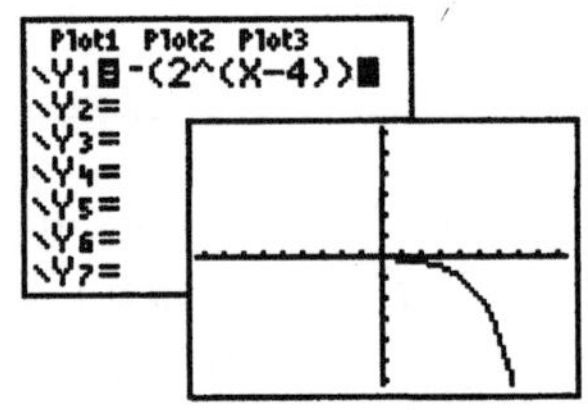

Observations: Replacing *x* by *x* - 4 shifts the graph four units. Multiplying the entire base by -1 reflects the graph about the *x*-axis.

EXERCISE SET

Directions: Match each graph to the appropriate exponential function. Use the calculator to check your results.

_______ 1. 5^x _______ 2. 5^{-x} _______ 3. $5^x - 1$ _______ 4. 5^{x-1}

_______ 5. -5^x _______ 6. $1 - 5^x$

A. 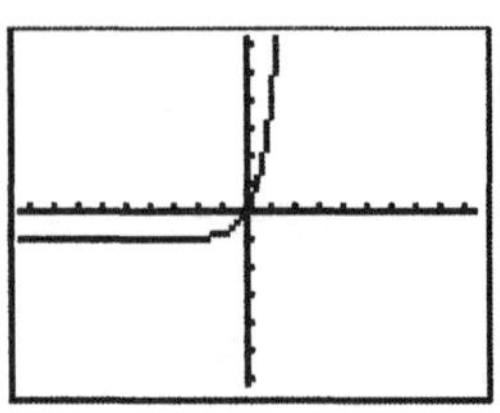

B. 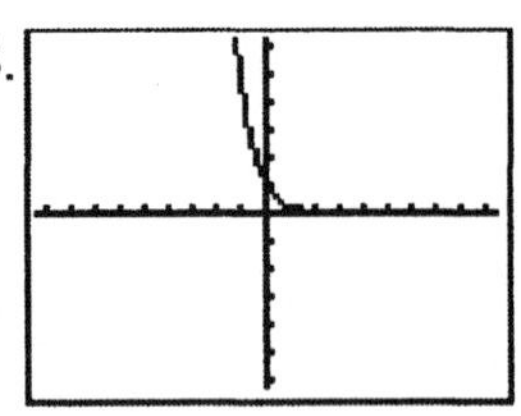

C. 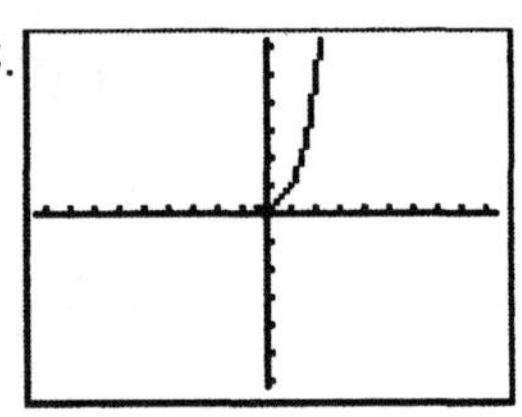

D. 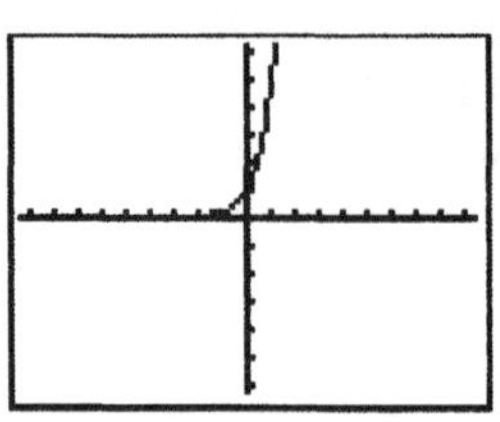

E. 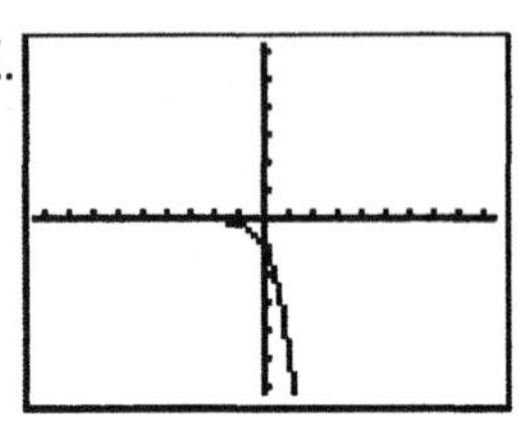

F. 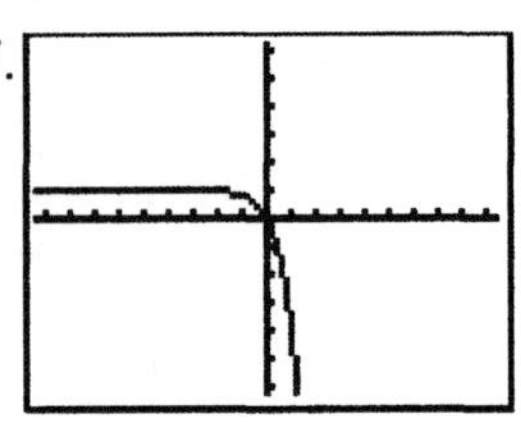

Base e Exponential Functions

The exponential function $f(x) = e^x$, with base *e*, an irrational number, occurs in natural applications. This constant (*e*) is located on the face of the calculator on the left side above the key labeled LN.

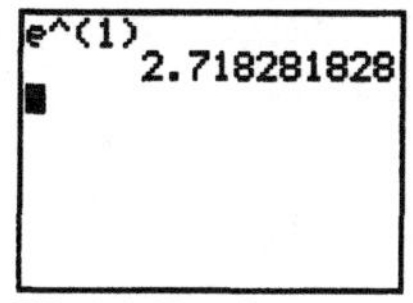

TI-83/84 Series

TI-86 Screen

EXERCISE SET CONT'D

✍7. Graph $y = 2^X$, $y = e^X$ and $y = 3^X$ on the screen at the right. Why does the graph of $y = e^X$ rise faster than the graph of $y = 2^X$ but not as fast as the graph of $y = 3^X$?

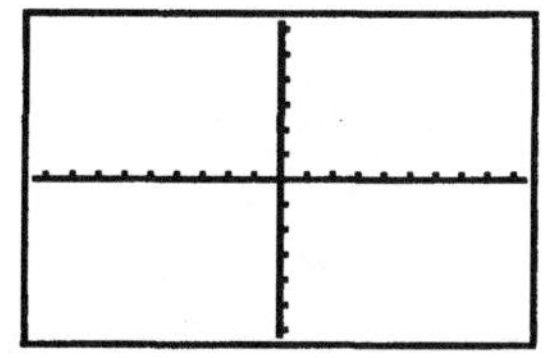

A common occurrence of the use of **e** in mathematical modeling is in the computation of compound interest, specifically continuous compounding. When an amount of money invested grows exponentially, the formula for computing the value of the investment is $A = Pe^{rt}$ where interest is compounded continuously.

8. If an initial investment of $2000 is placed in an account earning 4.5% interest compounded continuously, write an equation that will model the value of the investment at the end of *x* years.

____

9. Graph the equation from #8 on the screen at the right.
Viewing window hint: Since *x* represents the number of years, set the Xmax and Xmin to display non-negative values of *x*. To make the interpretation of the graphical display friendlier for tracing values, you may want to use the space width formulas discussed in the unit entitled *Calculator Viewing Windows* Remember, the initial investment is $2000; this will affect the Ymax.

10. TRACE along the graph and determine the value of the investment (to the nearest dollar) after

1 year________ 2 years ________ 5 years ________ 10 years ________

✍11. Explain how to use the TABLE to determine the value of the investment after 3 years 3 months. Record this value: ________

Logarithmic Functions

The inverse of an exponential function is a logarithmic function. To find the inverse of the exponential function ($y = b^x$) interchange the *x* and *y* variables. This yields the equation $x = b^y$ ($x > 0$ $b > 0, b \neq 1$) which is defined as the logarithmic function $y = \log_b x$. Until your text addresses Properties of Logarithms you will not have an algebraic method for solving this equation for *y*. Until that time, the **DrawInv** option from the DRAW menu will be used to draw the inverses of these functions.

Exponential Form	Graph	Graph of the Inverse	Logarithmic Form
$f(x) = 2^x$	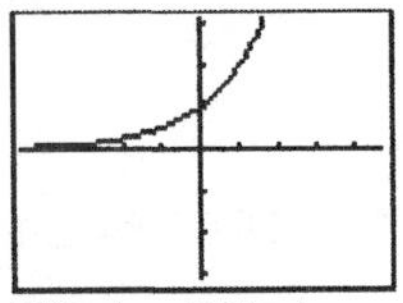 ZDecimal Window	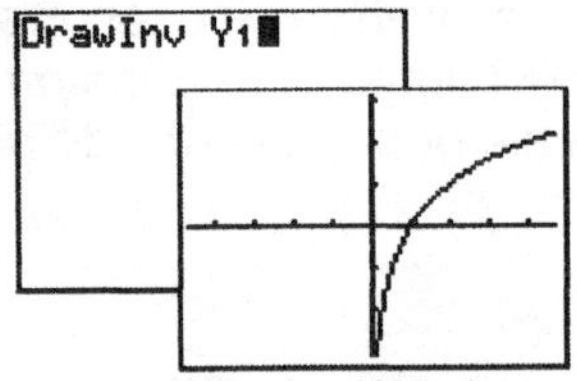 ZDecimal Window	$f(x) = \log_2 x$

Observations (Inverse Function): The x-intercept of the graph is 1; there is no y-intercept. The y-axis is a vertical asymptote.
The domain of the exponential function, $(-\infty, \infty)$, is the range of the logarithmic function.
The range of the exponential function, $(0, \infty)$, is the domain of the logarithmic function.

When the base of a logarithmic function is e we have the natural logarithm function denoted $f(x) = \ln x$.

Transformation of Logarithmic Functions

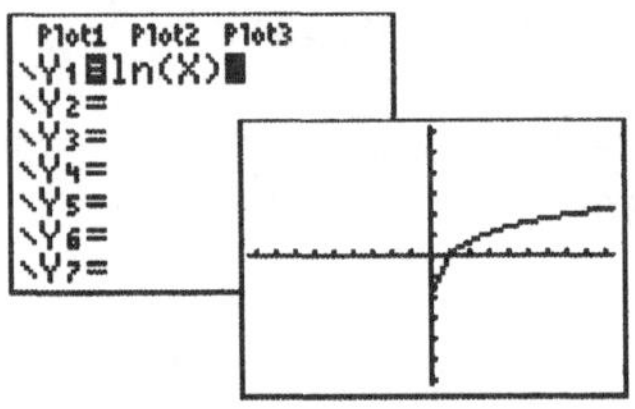	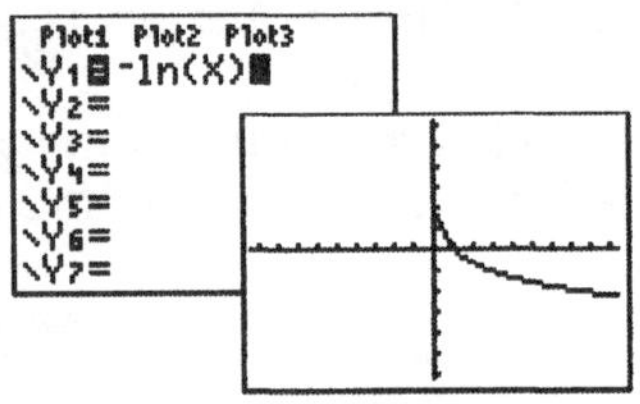	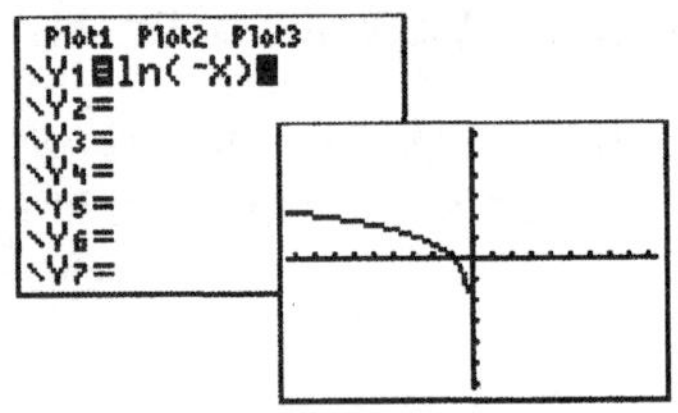

Observations: Multiplying by -1 reflects the graph across the x-axis whereas replacing x with $-x$ reflects the graph across the y-axis.

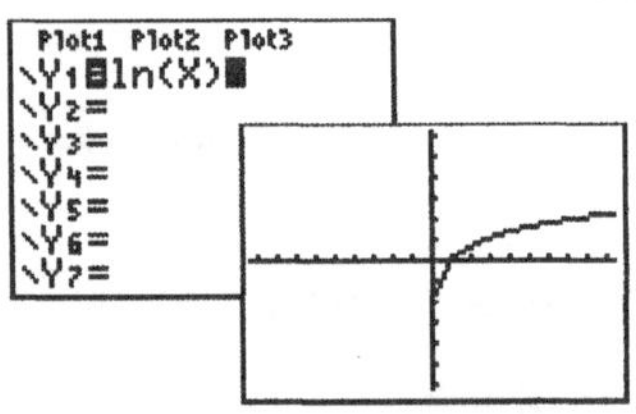	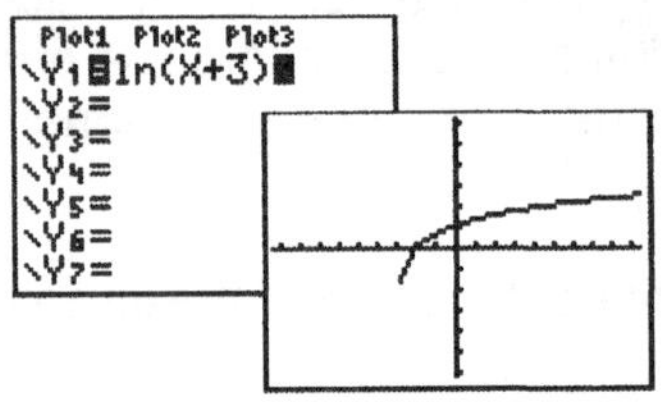	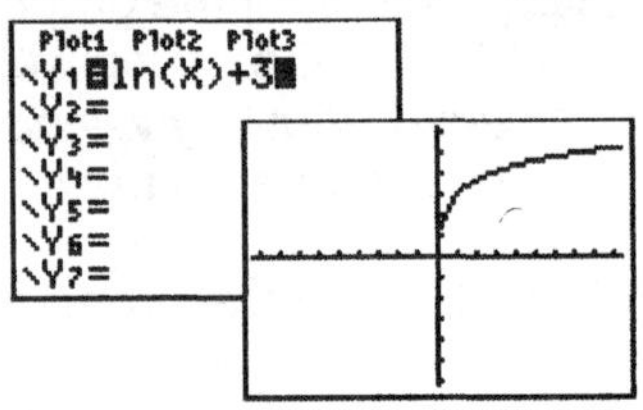

Observations: Replacing x with $x + 3$ shifts the graph *left* three units whereas adding 3 shifts the graph *up* three units. Comparable shifts occur with the operation of subtraction.

The graphing calculator has a key labeled LN (for natural log) as well as a key labeled LOG. The LOG key is defined for a base of10 logarithm (a common logarithm).

Common Logarithm	Graphing Calculator	Natural Logarithm	Graphing Calculator
$\log 100 = 2$ means $\log_{10}100=2$ which can be written in exponential form as $10^2 = 100$	log(100) 2 10^2 100	$\ln 100 \approx 4.6052$ means $\ln_e \approx 4.6052$ which can be written in exponential form as $e^{4.6052} \approx 100$	ln(100) 4.605170186 e^(Ans) 100 e^(4.6052) 100.0029814

A change of base formula is required to use the calculator to evaluate expressions containing logarithms with bases other than 10 or *e*.

Change of Base Formula: $\log_b n = \frac{\log n}{\log b} = \frac{\ln n}{\ln b}$.

Evaluate: $\log_2 5$	**Change of Base Formula:**	**Graphing Calculator:**
	$\log_b n = \frac{\log n}{\log b} = \frac{\log 5}{\log 2}$ or $\frac{\ln 5}{\ln 2}$	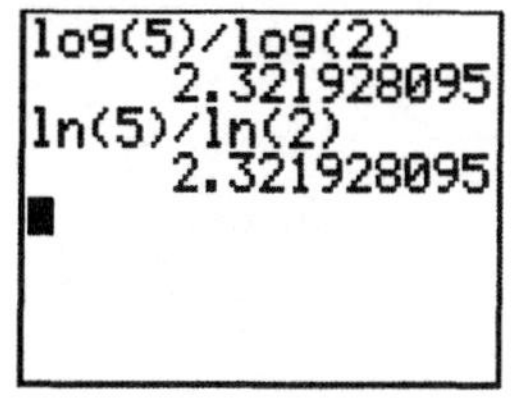

EXERCISE SET CONT'D

Directions: Use the change of base formula (if necessary) to correctly match each logarithmic expression to its calculator evaluation.

_______ 12. $\log_2 8$ _______ 13. log 8 _______ 14. log 2 _______ 15. $\log_8 2$

A. 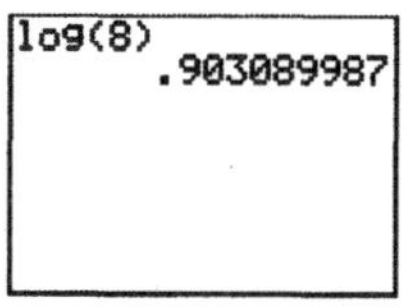

B. 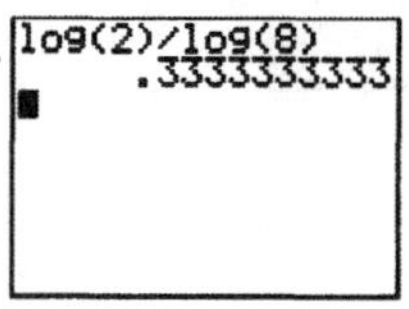

C. 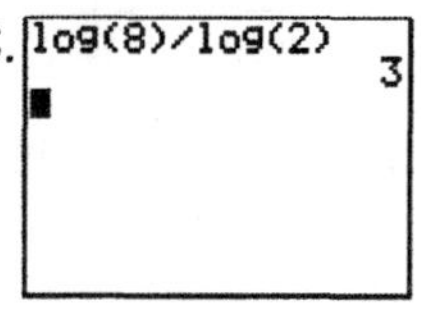

D. 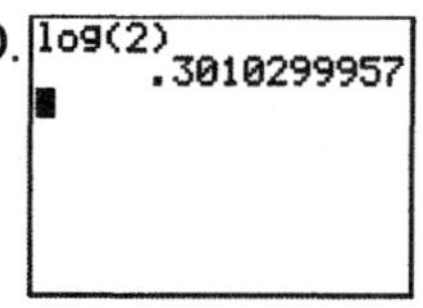

Exponential and Logarithmic Equations

Example	**Graphical Solution**	
Solve: $3^{x+1} = 27$	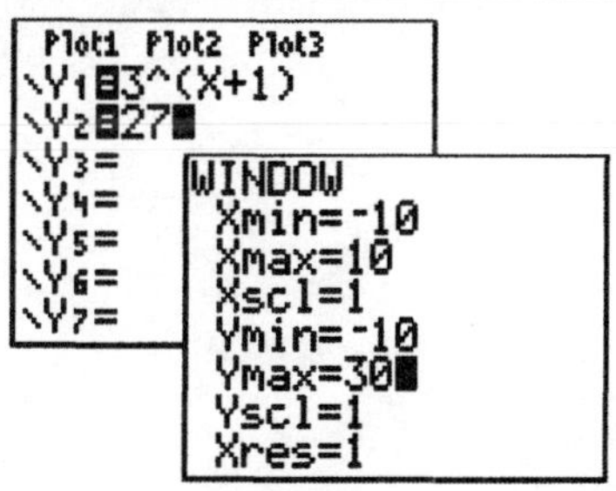	Enter the left side of the equation at Y1 and the right side at Y2. Adjust the viewing window so that the *y*-value of 27 is visible.
	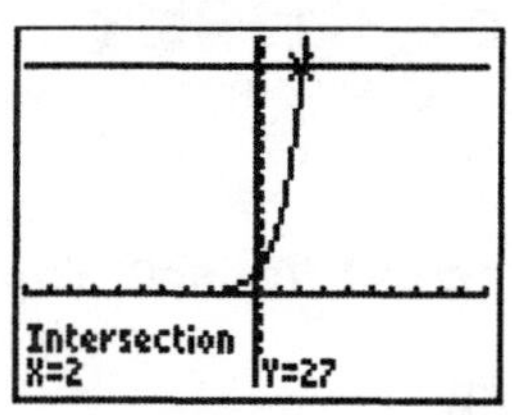	Use the INTERSECT feature of the calculator. The solution is 2.

<table>
<tr>
<td rowspan="2">Solve: $\log_2 (4x - 2) = 3$</td>
<td>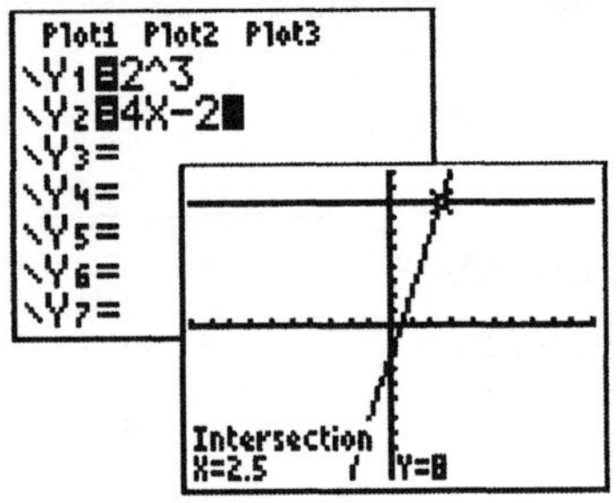
</td>
<td>Translate $\log_2 (4x - 2) = 3$ to exponential form and enter the left side at Y1 and the right side at Y2. Use the INTERSECT feature to find the solution.</td>
</tr>
<tr>
<td>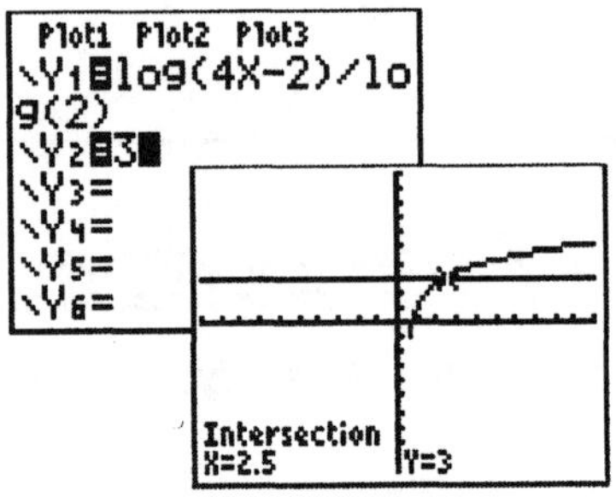
</td>
<td>Use the Change of Base formula on the left side of the equation, enter that expression at Y1 and the right side at Y2. Use the INTERSECT feature to find the solution.</td>
</tr>
</table>

EXERCISE SET CONT'D

Directions: Solve each of the equations below graphically.

16. $9 = 3^{1/2\,x+5}$

17. $e^{2x} = e^{x+1}$

18. $\log_4 64 = x$

19. $\log_x \frac{9}{4} = 2$

20. $\log_{1/2} 16 = x$

Solutions: **1.** D **2.** B **3.** A **4.** C **5.** E **6.** F

8. $y = 2000e^{.045x}$

9. Suggested window: Xmin=0, Xmax=94, Xscl=0, Ymin=0, Ymax=5000, Yscl=0

10. 1yr=2092, 2yr=2188, 5yr=2505, 10yr=3137

11. The TABLE will need to be incremented by .01 since 3 yrs. 3 mos. is 3.25 years. The value after 3.25 yrs. is $2315.

12. C **13.** A **14.** D **15.** B **16.** -6 **17.** 1 **18.** 3

19. 3/2 **20.** - 4

UNIT 21
CURVE OF BEST FIT

*Prerequisite: Unit #8.

This unit examines the relationship between two variable quantities (data points) in different types of application problems. We will use scatter plots to obtain a general idea of the shape of the data and then fit the data to the graph which exhibits the same behavior as the scatter plot. Once the shape of the curve has been selected, linear, quadratic, or cubic, the best fit will be determined by the correlation coefficient. The best model will yield a correlation value of *r* for which $|r|$ is closest to 1. The correlation value determines the strength of the relationship between the two variables; the closer the value to one - the stronger the relationship (note - the correlation value is noted as *corr* on the TI-86).

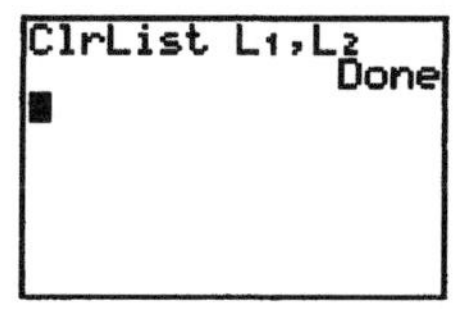

Before beginning, you will need to clear all data from the lists in the calculator and delete the entries from the *y*-edit screen. There are several approaches to clearing lists. For demonstration purposes, suppose that only the first and second lists (L1 and L2) are to be cleared. To clear these lists, from the home screen, press **[STAT] [4](4:ClrList) [2nd] <L1> [,] [2nd] <L2> [ENTER]**.
Another approach is to press **[STAT] [1](1:Edit),** use the ▲ key to highlight L1 and press **[CLEAR]**. Cursor down below L1 to see the list clear.

TI-86	PRESS **[2ND] <STAT> [F2] (EDIT)** USE THE CURSOR KEYS TO HIGHLIGHT **XSTAT** AND PRESS **[CLEAR] [ENTER]**. REPEAT THE PROCESS TO CLEAR THE **YSTAT** LIST.

These approaches are useful for clearing specific lists while leaving other lists intact. If the intent is to clear **all** lists then the quickest approach is to press **[2nd] [MEM] <4>(4:ClrAllLists) [ENTER].**

TI-86	GO TO PAGE 133.

Scatter Plots

Scatter Plots	Calculator Display	
Use a graphing calculator to find a scatter plot data points: (-5, -4), (-1,0), (0,1),(1,4), (2,5)	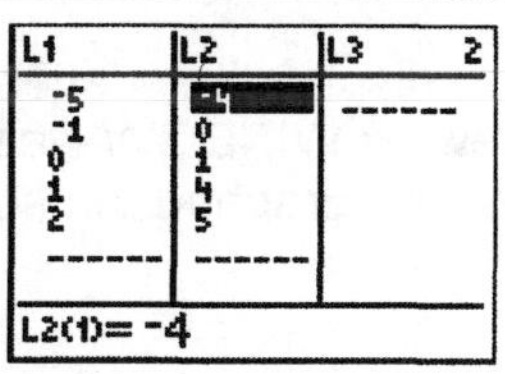 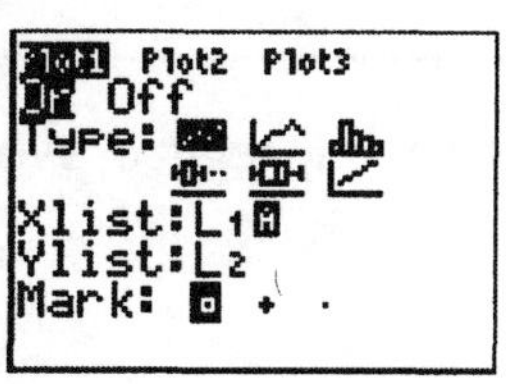 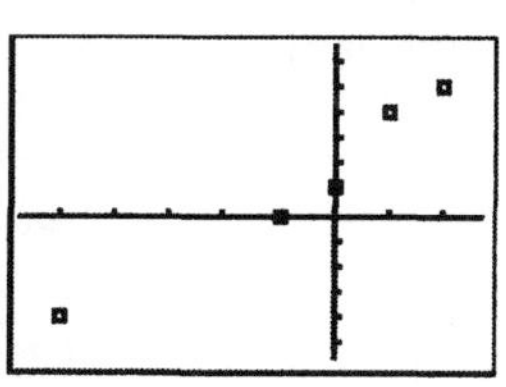	•Press **[STAT] [1](1:Edit)** to enter paired data at L1 and L2. Enter *x*-values, in the L1 list, pressing *enter* after each entry. Use the ▸ arrow key to cursor to the L2 list and enter the corresponding *y*-values. Be sure data values correspond to the originals. •Press **[2nd] <STATPLOT> [1](1:Plot 1)**. Press **[ENTER]** to turn on the first plot. Use the ▾ arrow key to highlight the first icon after **Type** and press **[ENTER]**. Use the ▾ arrow key followed by **[2nd]<L1>** to enter **L1** after **Xlist** if L1 is not already displayed. Then use **[▾]** to **Ylist** and press **[2nd] <L2>** to display **L2** after **Ylist**. Finally, use the ▾to highlight the first entry after **Mark** and press **[ENTER]**. Plotted data points can be represented by □, +, or a •. We selected □ for easy visibility. •Press **[ZOOM]** and cursor down and press **[9](9:ZoomStat)** to view the scatter plot.

We will now use the scatter plot example above to examine the curve of best fit. The initial steps for finding the curve of best fit are the same as for determining a scatter plot. They are only briefly restated in this next example.

Curve of Best Fit: Linear	**Graphical Display**	
Use a graphing calculator to find a curve of best fit for the data points: (-5, -4), (-1,0), (0,1), (1,4), (2,5)	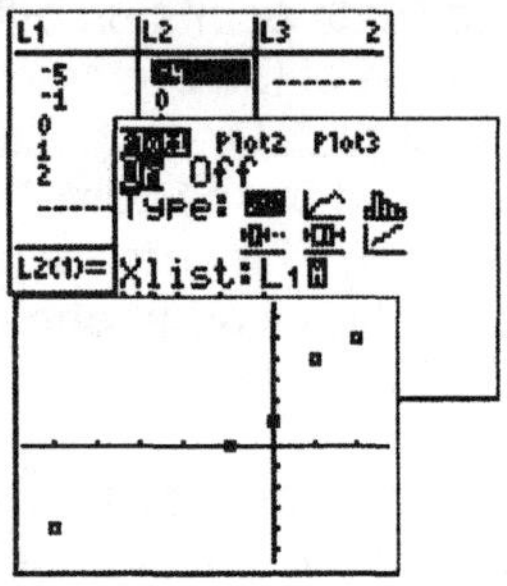 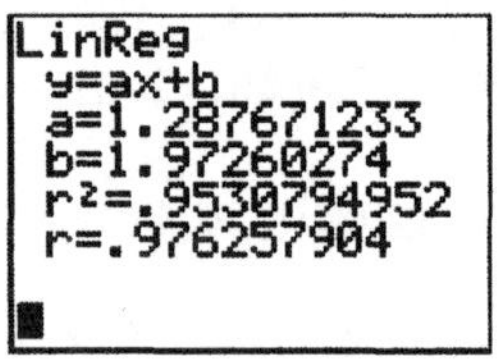 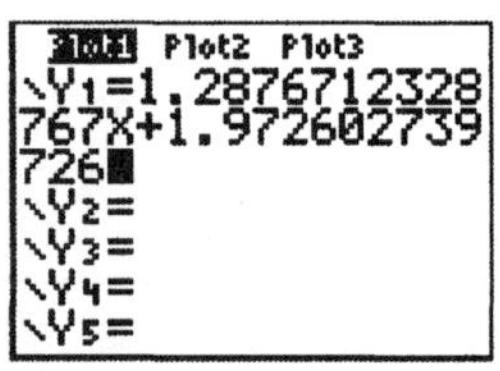 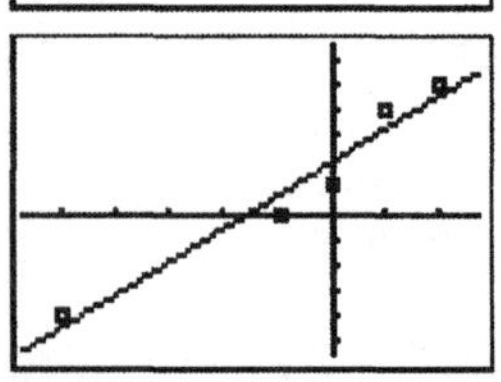	•Enter data points at L1 and L2.(Note that these are the same data points as used in the scatter plot thus the first 3 steps will have already been completed.) •Press **[2nd] <STATPLOT>** and turn on the desired statplot. •View the scatter plot in the zoomstat window **and decided what type if curve best describes this plot.** •Return to the home screen by pressing **[2nd]<QUIT>**. • Press **[STAT] [▸]**(to highlight CALC) **[4] (4:LinReg(ax+b)) [ENTER]** at the home screen. On the home screen the regression equation, and correlation coefficient, will be displayed. The correlation coefficient is not automatically displayed on the TI-83/84 series because the calculator default mode is set to **Diagnostic Off**. One way to determine the **r** value is to press **[VARS] [5] (5:Statistics) [▸] [▸]** to highlight **EQ**, and **[7](7:r) [ENTER]**. Or, if the r-value is always needed then the calculator should be reset to **Diagnostic On**: press **[2nd] <CATALOG> [▾]** until the pointer marks **Diagnostic On**, **[ENTER]** to copy the command to the home screen and **[ENTER]** again to activate the command. •To paste the regression equation to the y-edit screen, from Y1=, press **[VARS] [5](5:Statistics)[▸] [▸]** (to highlight EQ for equation) **[1](1: RegEQ)**. •Press **[GRAPH]** to display the scatter plot and the "curve of best fit". The viewing WINDOW was automatically set to ZoomStat when we first graphed the scatter plot.

When the curve of best fit is linear we say that the data points are linearly related. We will now consider some nonlinear relationships.

Curve of Best Fit: Quadratic

A quadratic relationship is determined by a minimum of 3 data points. If you have three data points then you have a polynomial fit. If you have four or more points you have a polynomial regression. At least 3 points are required. Be sure to clear the *y*-edit screen and the lists before beginning.

<table>
<tr><th>Curve of Best Fit: Quadratic</th><th>Calculator Display</th><th></th></tr>
<tr>
<td>Use a graphing calculator to find a curve of best fit for the data points: (-5,-8),(-4,1), (-3,5),(-1,11), (1.5,6), (2.5,0), (4,-7),(0,11)</td>
<td>
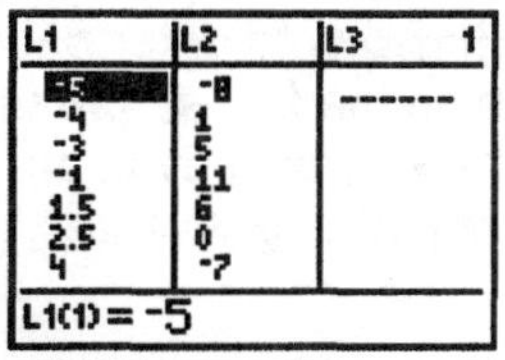

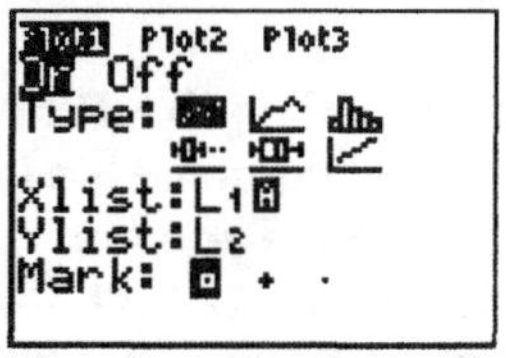

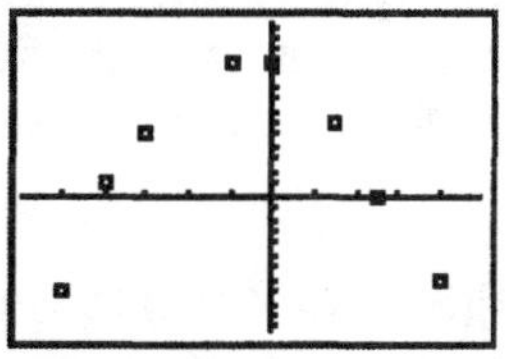
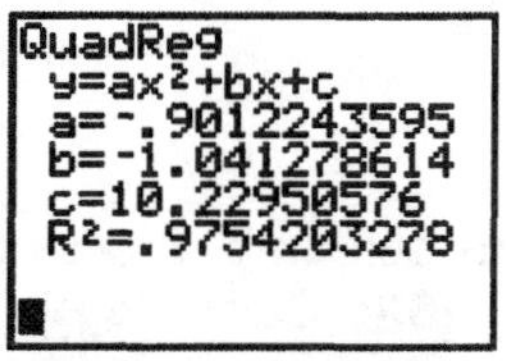

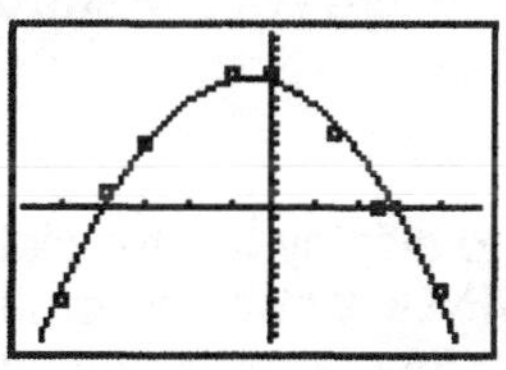
</td>
<td>
•Press [STAT] [1:](1Edit) to enter paired data at L1 and L2. Enter x-values, in the L1 list, pressing enter after each entry. Use the ▸ arrow key to cursor to the L2 list and enter the corresponding y-values.

•Press [2nd] <STATPLOT> [1](1:Plot 1). Press [ENTER] to turn on the first plot. Use the ▾ arrow key to highlight the first icon after Type and press [ENTER].Use the ▾ arrow key followed by [2nd]<L1> to enter L1 after Xlist if L1 is not already displayed. Then use [▾] to Ylist and press [2nd] <L2> to display L2 after Ylist. Finally, use the ▾to highlight the first entry after Mark and press [ENTER].

•Press [ZOOM] [9](9:ZoomStat) to view the scatter plot.

•Return to the home screen by pressing [2nd]<QUIT>.

• Press [STAT] [▸] (to highlight CALC) [5](5:QuadReg)) [ENTER] at the home screen.

•To paste the regression equation to the y-edit screen, from Y1=, press [VARS] [5](5:Statistics)[▸] [▸] (to highlight EQ for equation) [1](1: RegEQ).

•Press [GRAPH] to display scatter plot and curve of best fit. The viewing WINDOW was automatically set to ZoomStat when we first graphed the scatter plot.
</td>
</tr>
</table>

Curve of Best Fit: Cubic

A cubic relationship is determined by a minimum of 4 data points. If you have four data points then you have a polynomial fit. If you have five or more points you have a polynomial regression. At least 4 points are required. Be sure to clear the *y*-edit screen and the lists before beginning.

Curve of Best Fit: Cubic	**Calculator Display**	
Use a graphing calculator to find a curve of best fit for the data points: (2,650), (3,1006), (4,1055), (6,835), (8,442), (9,130), (10, -110), (11,-350), (12,-289), (13,-250), (14,-260), (16, 906), (17,1000)	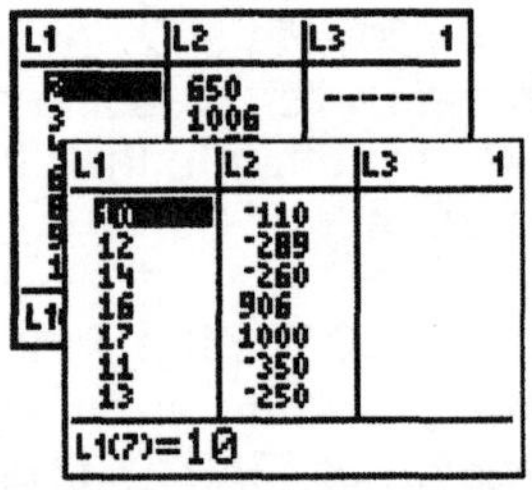	•Press **[STAT] [1:](1Edit)** to enter paired data at L1 and L2. Enter *x*-values, in the L1 list, pressing *enter* after each entry. Use the ▸ arrow key to cursor to the L2 list and enter the corresponding *y*-values.
	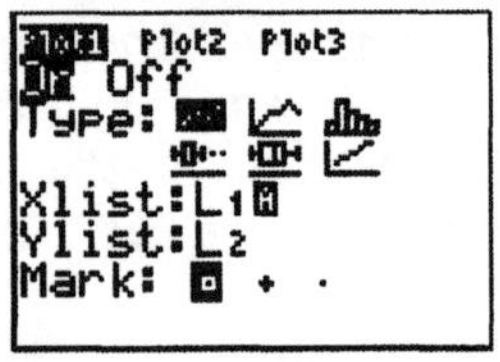	•Press **[2nd] <STATPLOT> [1](1:Plot 1)**. Press **[ENTER]** to turn on the first plot. Use the ▾ arrow key to highlight the first icon after **Type** and press **[ENTER]**.Use the ▾ arrow key followed by **[2nd]<L1>** to enter **L1** after **Xlist** if L1 is not already displayed. Then use **[▾]** to **Ylist** and press **[2nd] <L2>** to display **L2** after **Ylist**. Finally, use the ▾to highlight the first entry after **Mark** and press **[ENTER]**.
	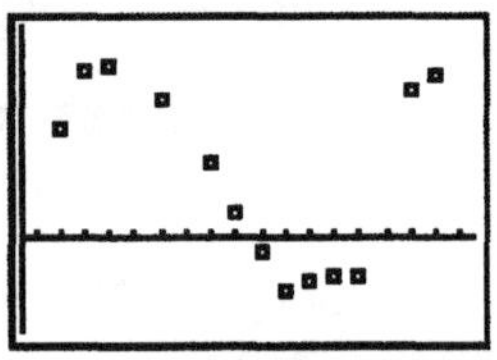	•Press **[ZOOM] [9](9:ZoomStat)** to view the scatter plot.
	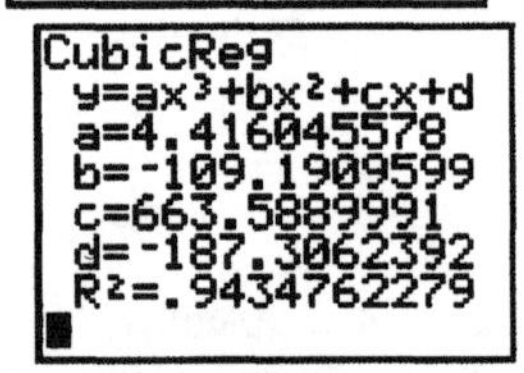	•Return to the home screen by pressing **[2nd]<QUIT>**. • Press **[STAT] [▸] [6](6:CubicReg)) [ENTER]** at the home screen
		•To paste the regression equation to the y-edit screen, from Y1=, press **[VARS] [5](5:Statistics)[▸] [▸]** (to highlight EQ for equation) **[1](1: RegEQ)**.
	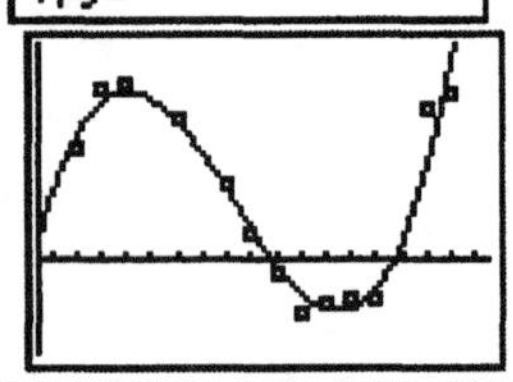	•Press **[GRAPH]** to display scatter plot and curve of best fit. The viewing WINDOW was automatically set to ZoomStat when we first graphed the scatter plot.

Summary: Curve of Best Fit

- **Clear *y*-edit and lists.**
- **Enter data.**
- **Turn plots ON and choose the desired display type.**
- **Set viewing window.**
- **On the home screen display the regression equation.**
- **Paste that equation to the *y*-edit screen.**
- **Display the graph.**

EXERCISE SET

1. The chart below gives the year and winning time (in seconds) for the Men's 1000-Meter Speed Skating event. Display a scatter plot of these data points in an appropriate viewing window.

YEAR	1976	1980	1984	1988	1992	1994	1998
TIME	79.32	75.18	75.8	73.03	74.85	72.43	70.64

(Source: *The New York Times 1998 Almanac*))

Graph display:

2. The following data is based on the Tennessee Department of Education annual report card of the Clarksville-Montgomery County School System. Determine a scatter plot for the data and a line of best fit for the data.

Year	1993	1994	1995	1996	1997	1998	1999	2000	2001	2002
Hispanic Student Enrollment	481	602	629	709	774	885	956	979	1035	1116

Graph of scatter plot and regression line

3. A survey was taken of notably tall buildings on the east coast, mid-west, and west coast. The survey compares height, in feet from sidewalk to roof, to number of stories (beginning at street level). If the Empire State building has 102 stories, use the given information to predict the height of the building: display a scatter plot, determine the linear regression equation, use the calculator table to predict the height.
(Source: The World Almanac and Book of Facts 1995, actual height is 1250 ft.)

BUILDING	STORIES	HEIGHT (FT.)
Baltimore U.S. Fidelity and Guaranty Co.	40	529
Maryland National Bank (Baltimore, MD)	34	509
Sears Tower (Chicago, IL)	110	1454
John Hancock Building (Chicago, IL)	100	1127
Transamerica Pyramid (San Francisco, CA)	48	853
Bank of America (San Francisco, CA)	52	778

Predicted height of the Empire State Building:_____

Value (to the nearest hundredth) of the correlation coefficient?______

Graph of scatter plot and regression line

4.. The following table gives the winning time (in seconds) for the Men's 400-Meter Freestyle swimming event in the Olympics from 1924 to 1956. The Olympics were not held during the years 1916, 1940, and 1944 due to the two World Wars. Find a linear regression equation to predict the winning times for each of those years.

YEAR	1924	1928	1932	1936	1948	1952	1956
TIME	304.2	301.6	288.4	284.5	281	270.7	267.3

(Source: *The New York Times 1998 Almanac*)

Predicted winning times: 1916 _______ 1940 _______

1944 ________

Value of the correlation coefficient (to the nearest hundredth): _______

Graph of scatter plot and regression line

Directions: On the following group of data points, determine the curve of best fit and the correlation coefficient.. Remember, the best model will yield a correlation value of *r* for which $|r|$ is closest to 1. Sketch the scatter plot and the curve

5.

x	10	20	30	40	50	60	70
y	60	61	61	62	63	64	65

Correlation coefficient: _____

6.

x	0	1	2	3	4	5
y	11	2	-4	-5	-3	1

Correlation coefficient:_______

7.

x	1	2	4	6	8	9	10
y	-3	-1	1	-1	1	4	8

Correlation coefficient:_______

8.

x	1	0	0	-1	-2	-3	-4	-4
y	-4	0	1	3	4	3	2	0

Correlation coefficient: _______

9.

x	1	2	3	4	5	6	7	8	9	10	11
y	1300	1400	1550	1600	1600	1600	1550	1500	1450	1500	1600

Correlation coefficient:_______

10.

x	10	13	15	17	18	20	23	24
y	144	153.5	164	161	162	180	179	190

Correlation coefficient: _______

Solutions:

1.

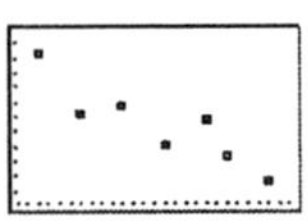

2.

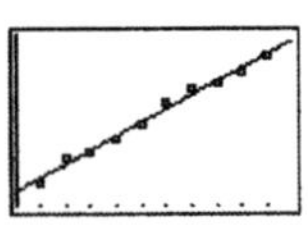

3. height: 1281.1, r≈.95

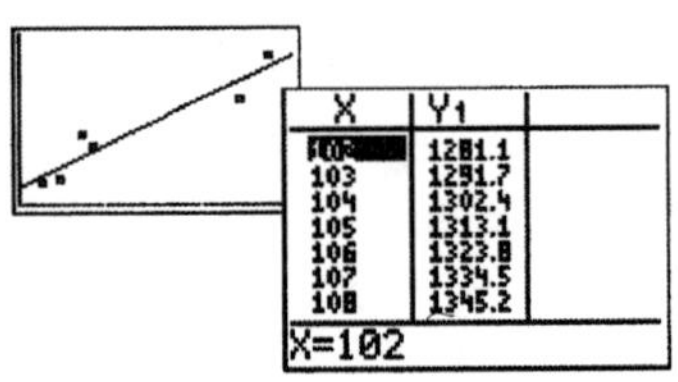

4. r ≈ -.96, 1916: 310.7
1940: 284.8, 1944: 280.5

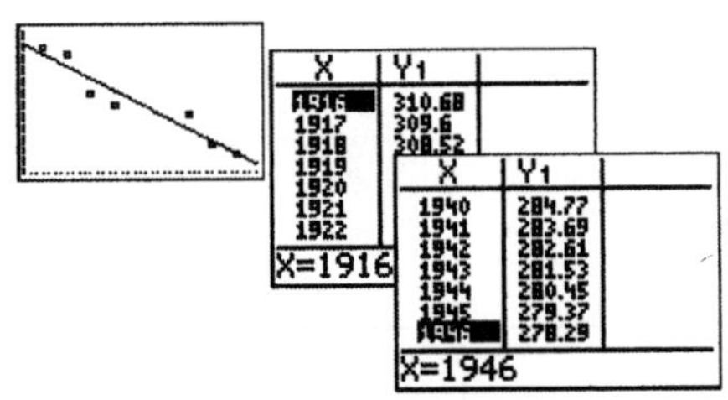

5. r≈ .99

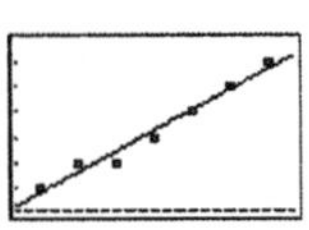

6. r^2≈ .99

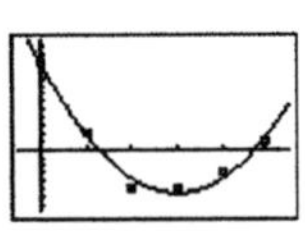

7. r^2≈ .99

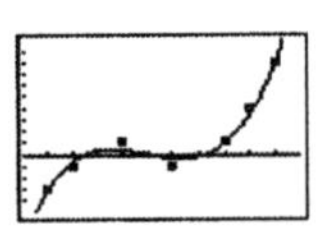

8. r^2≈.94

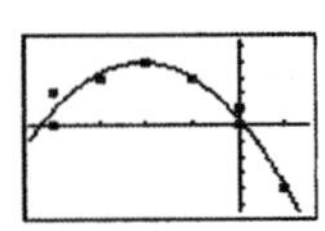

9. r^2 ≈ .94

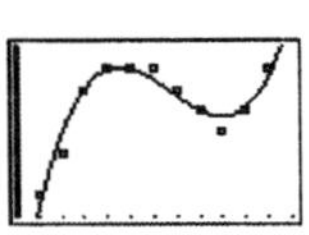

10. r ≈ .95

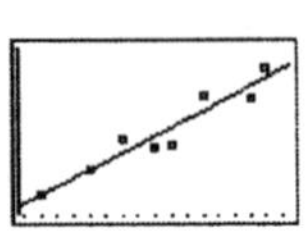

Scatter Plots

Scatter Plots	Calculator Display	
Use a graphing calculator to find a scatter plot for the data points: (-5, -4), (-1,0), (0,1),(1,4), (2,5)	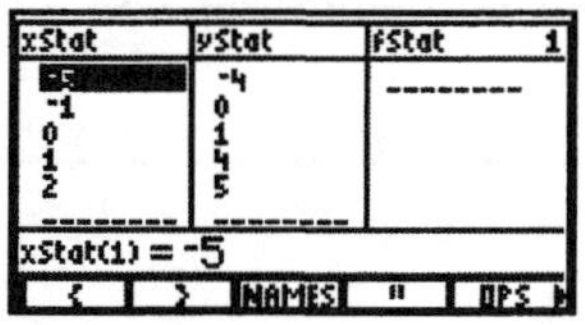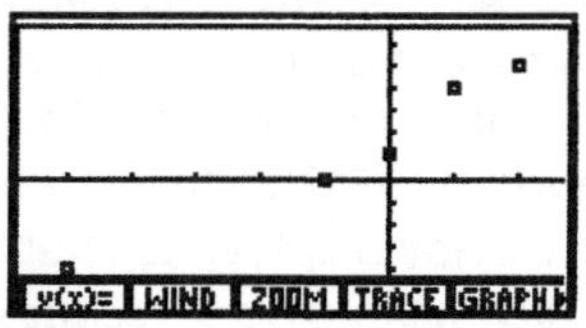	•Press **[2nd] <STAT>** and **[F2] (EDIT)** to enter the paired data at xStat and yStat. Enter *x*-values, in the xStat list, pressing *enter* after each entry. Use the ▸ arrow key to cursor to the yStat list to enter corresponding *y*-values. •Press **[2nd] <STAT> [F3](PLOT)** and **[F1](PLOT1))** Press **[ENTER]** to turn on the first plot. Use the ▾ arrow key to move to **Type** and press **[F1](SCAT)**. Use the ▾ arrow key followedby **[F1](xStat)** to enter **xStat** after **Xlist** if it is not already displayed. Then use **[▾]** to **Ylist** and press **[F2](yStat)** to display **yStat** after **Ylist**. Finally, use the [▾] and press **[F1](□)** to enter □ after **Mark** . Plotted data points can be represented by □, +, or a •. We selected □ for easy visibility. •Press **[GRAPH] [F3] (ZOOM) [MORE] [F5] (ZDATA) [EXIT]**

Curve of Best Fit: Linear

We will now use the scatter plot example above to examine the curve of best fit. The initial steps are the same as for determining a scatter plot. They are only briefly restated in this next example

Curve of Best Fit: Linear	Graphical Display	
Use a graphing calculator to find a curve of best fit for the data points: (-5, -4), (-1,0), (0,1), (1,4), (2,5)	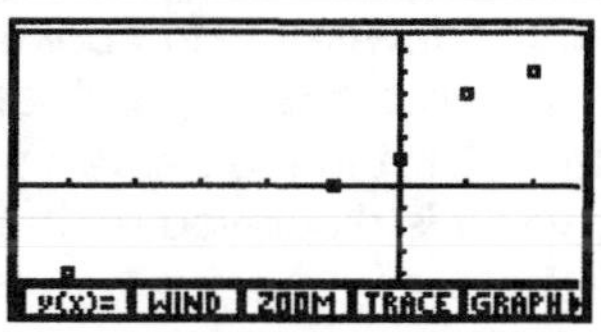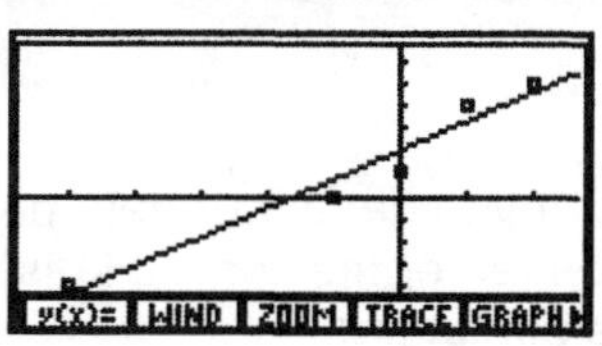	•Once the data is entered and you are looking at the display of your points, you should decide what curve will best fit the data display. *Note: the screens displayed are the same as above.* •Return to the home screen by pressing **[2nd] <QUIT>**. • Press **[2nd] <STAT> [F1] (CALC) [F3](LinR)** followed by **[2nd](CATLG/VARS) [F2](ALL)**. Cursor down until the *pointer* is at **xStat,** press **[ENTER]** then press **[,]** and return to the catalog to select **yStat** in the same manner. When *LinR xStat, yStat* is displayed on the home screen, press **[ENTER].** •Press **[GRAPH] [F1] (y(x))**.) To copy the linear regression equation to the **y(x)=** screen, press **[2nd] <CATLG-VARS> [MORE] [MORE] [F4] (STAT)**, cursor down to **RegEq** and press **[ENTER]**. •Press **[2nd] <M5> (GRAPH)** to view the scatterplot and the curve of best fit .The viewing window was set to ZDATA when we graphed the scatter plot.

When the curve of best fit is linear we say that the data points are linearly related. We will now consider some nonlinear relationships. Be sure to clear the *y*-edit screen and the lists before continuing.

Curve of Best Fit: Quadratic

A quadratic relationship is determined by a minimum of 3 data points. If you have three data points then you have a polynomial fit. If you have four or more points you have a polynomial regression. At least 3 points are required.

<table>
<tr><th>Curve of Best Fit: Quadratic</th><th>Calculator Display</th><th></th></tr>
<tr>
<td>Use a graphing calculator to find a curve of best fit for the data points: (-5,-8),(-4,1), (-3,5),(-1,11), (1.5,6), (2.5,0), (4,-7),(0,11)</td>
<td>
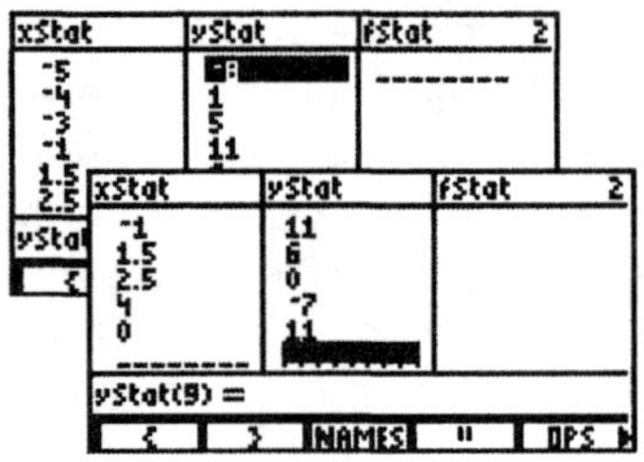

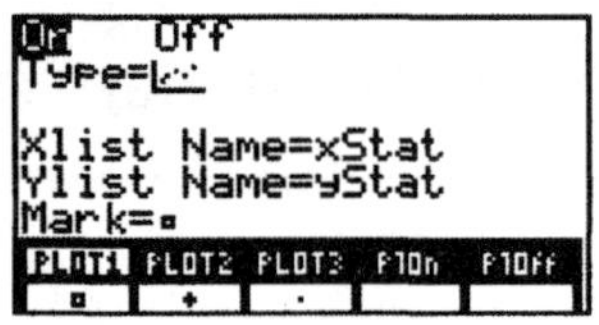

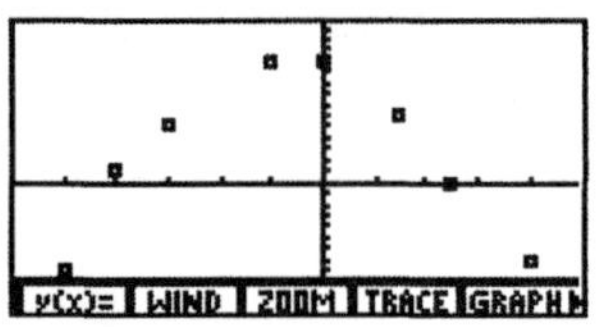

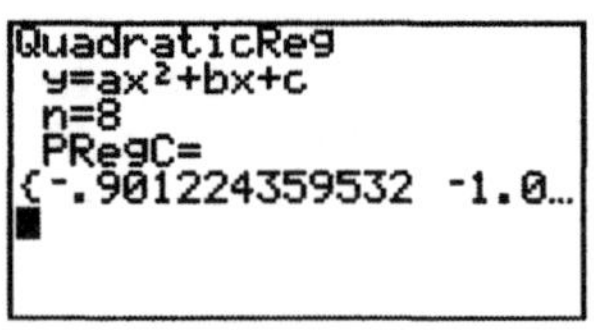

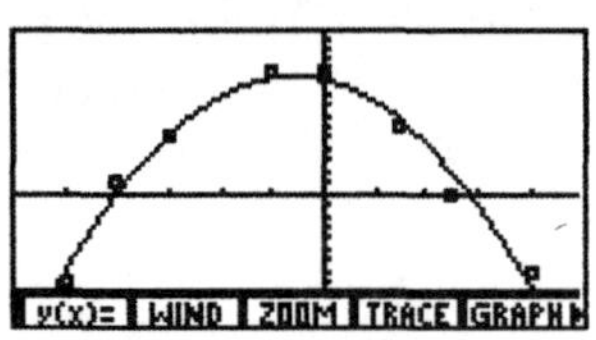

</td>
<td>•Press **[2nd] <STAT>** and **[F2] (EDIT)**. Enter *x*-values, in the xStat list, pressing *enter* after each entry. Use the ▸ arrow key to cursor to the yStat list and enter the corresponding *y*-values.

•Press **[2nd] <STAT> [F3](PLOT)** and **[F1](PLOT1))** Press **[ENTER]** to turn on the first plot. Use the ▾ arrow key to move to **Type** and press **[F1](SCAT)**. Use the ▾ arrow key followed by **[F1](xStat)** to enter **xStat** after **Xlist** if it is not already displayed. Then use **[▾]** to **Ylist** and press **[F2](yStat)** to display **yStat** after **Ylist**. Finally, use the [▾] and press **[F1](t □)** to enter □ after **Mark** .

•Press **[GRAPH] [F3] (ZOOM) [MORE] [F5] (ZDATA) [EXIT]**.

•Return to the home screen by pressing **[2nd]<QUIT>**.
• Press **[2nd] <STAT> [F1] (CALC) [MORE] [F4](P2Reg)** followed by **[2nd](CATLG/VARS) [F2](ALL)**. Cursor down until the *pointer* is at **xStat,** press **[ENTER]** then press **[,]** and return to the catalog to select **yStat** in the same manner. When *P2Reg xStat, yStat* is displayed on the home screen, press **[ENTER]**.)

•Press **[GRAPH] [F1] (y(x))**.) To copy the regression equation to the **y(x)=** screen, press **[2nd] <CATLG-VARS> [MORE] [MORE] [F4] (STAT)**, cursor down to **RegEq** and press **[ENTER]**.

•Press **[2nd] <M5> (GRAPH)** to view the scatter plot and the curve of best fit. The viewing WINDOW was automatically set to ZDATA when we first graphed the scatter plot.</td>
</tr>
</table>

Curve of Best Fit: Cubic

A cubic relationship is determined by a minimum of 4 data points. If you have four data points then you have a polynomial fit. If you have five or more points you have a polynomial regression. At least 4 points are required. . Be sure to clear the *y*-edit screen and the lists before continuing.

Curve of Best Fit: Cubic	**Calculator Display**	
Use a graphing calculator to find a curve of best fit for the data points: (2,650), (3,1006), (4,1055), (6,835), (8,442), (9,130), (10, -110), (11,-350), (12,-289), (13,-250), (14,-260), (16, 906), (17,1000)	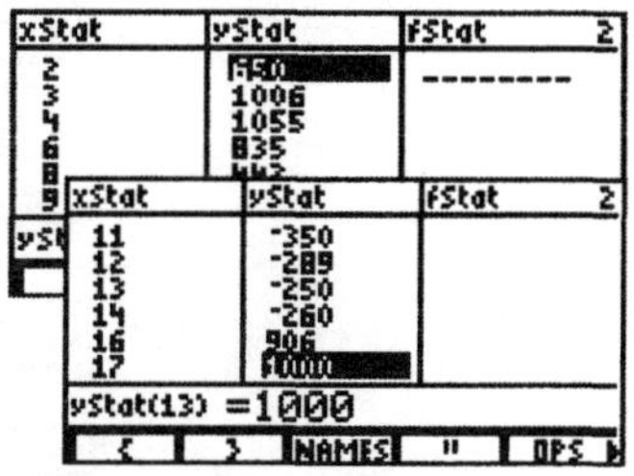	•Press **[2nd] <STAT>** and **[F2] (EDIT)**. Enter *x*-values, in the xStat list, pressing *enter* after each entry. Use the ▸ arrow key to cursor to the yStat list and enter the corresponding *y*-values.
		•Press **[2nd] <STAT> [F3](PLOT)** and **[F1](PLOT1))** Press **[ENTER]** to turn on the first plot. Use the ▾ arrow key to move to **Type** and press **[F1](SCAT)**. Use the ▾ arrow key followed by **[F1](xStat)** to enter **xStat** after **Xlist** if it is not already displayed. Then use **[▾]** to **Ylist** and press **[F2](yStat)** to display **yStat** after **Ylist**. Finally, use the [▾] and press **[F1](t □)** to enter □ after **Mark** .
	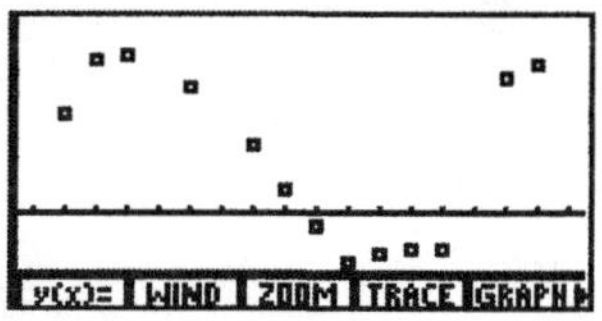	•Press **[GRAPH] [F3] (ZOOM) [MORE] [F5] (ZDATA) [EXIT]**
	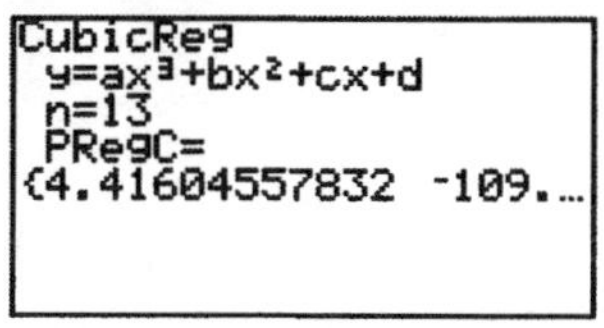	•Return to the home screen by pressing **[2nd] <QUIT>**. • Press **[2nd] <STAT> [F1] (CALC) [MORE] [F5](P3Reg)** followed by **[2nd](CATLG/VARS) [F3](ALL)**. Cursor down until the *pointer* is at **xStat,** press **[ENTER]** then press **[,]** and return to the catalog to select **yStat** in the same manner. When *P3Reg xStat, yStat* is displayed on the home screen, press **[ENTER].**)
		•Press **[GRAPH] [F1] (y(x))**.) To copy the regression equation to the **y(x)=** screen, press **[2nd] <CATLG-VARS> [MORE] [MORE] [F4] (STAT)**, cursor down to **RegEq** and press **[ENTER]**.
	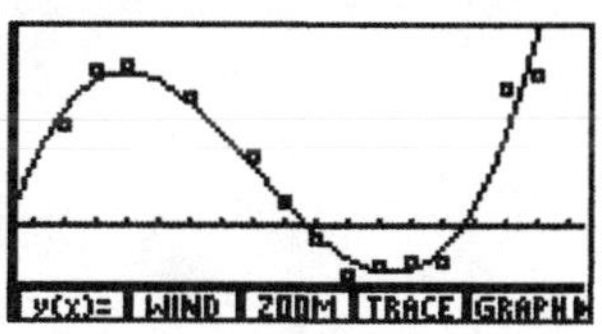	•Press **[2nd] <M5> (GRAPH)** to view the scatter plot and the curve of best fit. The viewing WINDOW was automatically set to ZDATA when we first graphed the scatter plot.

☞ RETURN TO THE SUMMARY STATEMENTS IN THE CORE UNIT (PG. 128).

UNIT 22
MATRICES

*Prerequisite: Unit # 3.

A matrix is a rectangular array of numbers. This unit will explore operations with matrices and their application to systems of linear equations.

To enter the matrix A = $\begin{bmatrix} 2 & 3 & 4 \\ 5 & 6 & 7 \end{bmatrix}$, press **[2nd] <MATRIX>** (on the TI-83, press **[MATRX]**) , cursor over to highlight **EDIT** and press **[ENTER]** to select **[1](1:[A])**.

> **TI-86** TO ENTER THE MATRIX A, PRESS **[2ND] <MATRX> [F2](EDIT)**. AT THE BLINKING ALPHA **CURSOR, TYPE <A> (TO NAME THE MATRIX) FOLLOWED BY [ENTER].**

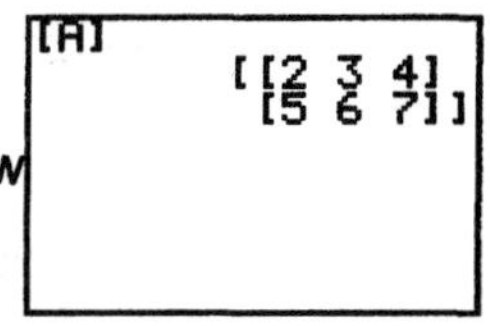

Matrix A has two rows and three columns. At the blinking cursor type **[2]**, press **[ENTER],** type **[3]**, and press **[ENTER]**. The values for the first row of the matrix can now be entered. Pressing **[ENTER]** after each entry will progress you through each row from left to right, or you may use the arrow keys to move to a desired location on the screen. Return to the home screen (**[2nd] <QUIT>**) and press **[2nd] <MATRIX> [1](1:[A]) [ENTER]** to display the matrix (remember, on the TI-83, just press **[MATRX]**, not **[2nd]**). Your display should correspond to the one at the right.

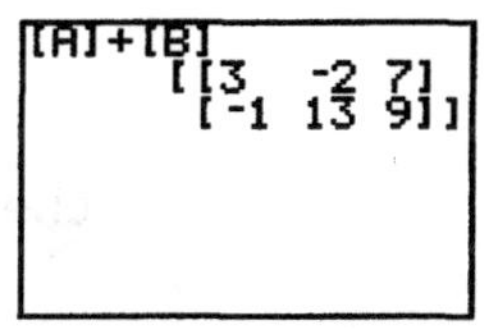

Now enter matrix B = $\begin{bmatrix} 1 & -5 & 3 \\ -6 & 7 & 2 \end{bmatrix}$ In order to add or subtract matrices, the dimensions of the matrices must be the same. At the home screen, press **[2nd] <MATRIX> [1](1:[A]) [+] [2nd] <MATRIX> [2](2:[B]) [ENTER].** Your display should look like the one at the right.

> **TI-86** PRESS **[2ND] <MATRX> [F1](NAMES) [F1](A) [+] [F2](B) [ENTER].** NOTE THAT THE MATRIX A IS SIMPLY DISPLAYED AS **A** AND NOT **[A]**.

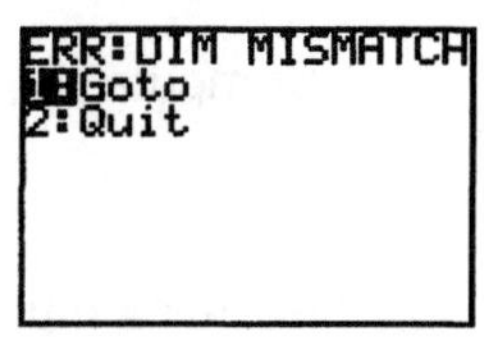

Now multiply: [A] · [B]. The screen display should correspond to the one at the right. In order to perform matrix multiplication, the dimensions must correspond as follows: Since [A] is a 2 x 3 matrix, it can only be multiplied by a 3 x n matrix.

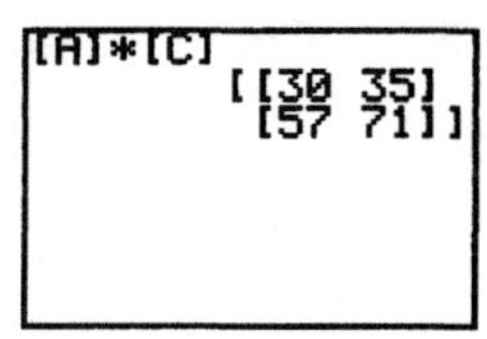

Enter the matrix C = $\begin{bmatrix} 4 & 5 \\ -2 & 3 \\ 7 & 4 \end{bmatrix}$ and multiply [A] · [C]. Your display should correspond to the one at the right.

EXERCISE SET

Directions: Enter the following matrices in the calculator:

$$A = \begin{bmatrix} 2 & 3 & 4 \\ 5 & 6 & 7 \end{bmatrix} \qquad B = \begin{bmatrix} 1 & -5 & 3 \\ -6 & 7 & 2 \end{bmatrix} \qquad C = \begin{bmatrix} 4 & 5 \\ -2 & 3 \\ 7 & 4 \end{bmatrix} \qquad D = \begin{bmatrix} 3 & 1 \\ -1 & 3 \end{bmatrix}$$

1. Using matrix A and matrix C, determine if matrix multiplication is a commutative operation.

✍2. What requirements are necessary for matrix multiplication to be defined?

3. Compute ([A] + [B])[C] and record the resulting matrix: $\begin{bmatrix} \_\_ & \_\_ \\ \_\_ & \_\_ \end{bmatrix}$

Compute [A][C] + [B][C] and record the resulting matrix: $\begin{bmatrix} \_\_ & \_\_ \\ \_\_ & \_\_ \end{bmatrix}$

Is matrix multiplication distributive from the right? ________

NOTE: If the name of a matrix is followed by an open parenthesis it <u>does</u> <u>not</u> indicate implied multiplication! A multiplication symbol must be used when parentheses <u>follow</u> a matrix name.

4 Compute [C]([A] + [B]) and record the resulting matrix:

This must be entered as [C] * ([A] + [B]). $\begin{bmatrix} \_\_ & \_\_ & \_\_ \\ \_\_ & \_\_ & \_\_ \\ \_\_ & \_\_ & \_\_ \end{bmatrix}$

Compute [C][A] + [C][B] and record the resulting matrix: $\begin{bmatrix} \_\_ & \_\_ & \_\_ \\ \_\_ & \_\_ & \_\_ \\ \_\_ & \_\_ & \_\_ \end{bmatrix}$

Is matrix multiplication distributive from the left? _______

✍5. Compute each of the following and record the resulting matrices:

[D]([A] + [C]) $\begin{bmatrix} \_\_ & \_\_ \\ \_\_ & \_\_ \end{bmatrix}$ [D][A] + [D][C] $\begin{bmatrix} \_\_ & \_\_ \\ \_\_ & \_\_ \end{bmatrix}$

In general, is matrix multiplication distributive?______

What requirement is necessary to ensure distributivity?

6. **Application**: The Golden Oldies: Records, Tapes and CDs chain has a warehouse in Kentucky and one in Tennessee. Use matrix addition and multiplication to determine the total value of the inventory for each "oldies" group that is listed. Each LP (long playing record) is valued at $7, each cassette tape at $9 and each CD (compact disc) is valued at $17.

Warehouse	**GROUP**	**LPs**	**TAPES**	**CDs**
KY Warehouse	"The Beagles"	125	250	275
	"Herman's Hideaways"	80	300	115
	"Peter, Piper, and Pepper"	75	185	200
TN Warehouse	"The Beagles"	200	180	200
	"Herman's Hideaways"	125	150	165
	Peter, Piper, and Pepper"	50	90	125

Record your matrix problem below, perform the indicated operation with your calculator and record your response to the problem.

The inverse of matrix A has the characteristic that $[A] \cdot [A]^{-1}$ is equal to the identity matrix. The multiplicative identity matrix is a square matrix that has ones along the diagonal (from upper left to lower right) and zeroes everywhere else. In the multiplication of real numbers, a number x has an inverse 1/x ($x \neq 0$) and the property that

$x \cdot 1/x = 1$ where 1 is the multiplicative identity. Multiplicative inverses exist only for some square matrices.

To compute the inverse of matrix A, first enter $A = \begin{bmatrix} 2 & 5 & 4 \\ 1 & 4 & 3 \\ 1 & -3 & -2 \end{bmatrix}$

At the home screen, enter A^{-1} by pressing **[2nd] <MATRIX> [1](1: [A]) [x^{-1}] [ENTER].** A^{-1} is displayed at the right.

TI-86	THE "x^{-1}" IS LOCATED ABOVE THE "EE" KEY ON THE FACE OF THE CALCULATOR.

$$A^{-1} = \begin{bmatrix} -1 & 2 & 1 \\ -5 & 8 & 2 \\ 7 & -11 & -3 \end{bmatrix}$$

NOTE: If this matrix needs to be stored for future reference, it can be stored in one of the other available matrix locations. To store A^{-1} in the matrix [E] location, type **[2nd] <MATRIX> [1: [A]] [STO▸] [2nd] < MATRIX> [5](5: [E]) [ENTER].**

Multiply A · A^{-1} to display the 3 x 3 identity matrix. This matrix should be numerically equivalent to the identity matrix displayed by pressing **[2nd] <MATRIX> [▸] [5] (5:identity) [3]**(for a 3 x 3 identity matrix) **[ENTER]**.

TI-86	THE IDENTITY FUNCTION IS FOUND UNDER THE **OPS** SUBMENU OF **MATRX, [2ND] <MATRX> [F4](OPS) [F3](IDENT).**

Because the inverse of A exists, A is called a nonsingular matrix. If the inverse of A did not exist, then A would be called a singular matrix.

If the inverse of a matrix exists, the multiplication of the matrix times its inverse is a commutative operation. (Test this idea using A and A^{-1}.)
Matrix multiplication can be used to rewrite linear systems in matrix form and then solved.

Example 1: Solve the linear system $\begin{matrix} x+4y=2 \\ 3x-2y=6 \end{matrix}$ $\begin{matrix} x+4y=2 \\ 3x-2y=6 \end{matrix}$ using matrices.

Solution: The linear system would be rewritten in matrix form as

$\begin{bmatrix} 1 & 4 \\ 3 & -2 \end{bmatrix} \bullet \begin{bmatrix} x \\ y \end{bmatrix} = \begin{bmatrix} 2 \\ 6 \end{bmatrix}$. Then proceeding with the matrix algebra:

$$\begin{bmatrix} 1 & 4 \\ 3 & -2 \end{bmatrix}^{-1} \bullet \begin{bmatrix} 1 & 4 \\ 3 & -2 \end{bmatrix} \bullet \begin{bmatrix} x \\ y \end{bmatrix} = \begin{bmatrix} 1 & 4 \\ 3 & -2 \end{bmatrix}^{-1} \bullet \begin{bmatrix} 2 \\ 6 \end{bmatrix}$$

$$I_2 \bullet \begin{bmatrix} x \\ y \end{bmatrix} = \begin{bmatrix} 1 & 4 \\ 3 & -2 \end{bmatrix}^{-1} \bullet \begin{bmatrix} 2 \\ 6 \end{bmatrix}$$

$$\begin{bmatrix} x \\ y \end{bmatrix} = \begin{bmatrix} 1 & 4 \\ 3 & -2 \end{bmatrix}^{-1} \bullet \begin{bmatrix} 2 \\ 6 \end{bmatrix}$$

Therefore, to solve a linear system with matrices, the matrix formed by the constants should be multiplied by the inverse of the coefficient matrix.

Enter A = $\begin{bmatrix} 1 & 4 \\ 3 & -2 \end{bmatrix}$, B = $\begin{bmatrix} 2 \\ 6 \end{bmatrix}$ and multiply A^{-1} · B.

The solution matrix A^{-1}· B is displayed at the right. $\begin{bmatrix} 2 \\ 0 \end{bmatrix}$

The solution to the linear system $\begin{matrix} x+4y=2 \\ 3x-2y=6 \end{matrix}$ is the ordered pair (2, 0).

◆

Example 2: Solve the system
$$\begin{aligned} x+\ y+z&=\ 7\\ x+2y+z&=-1\\ 2x+y+z&=\ 2 \end{aligned}$$
using matrices.

Solution: First, record in matrix form:

$$\begin{bmatrix} 1 & 1 & 1 \\ 1 & 2 & 1 \\ 2 & 1 & 1 \end{bmatrix} \bullet \begin{bmatrix} x \\ y \\ z \end{bmatrix} = \begin{bmatrix} 7 \\ -1 \\ 2 \end{bmatrix}$$

Using the MATRX EDIT screen, enter matrix A as $\begin{bmatrix} 1 & 1 & 1 \\ 1 & 2 & 1 \\ 2 & 1 & 1 \end{bmatrix}$ and matrix B as $\begin{bmatrix} 7 \\ -1 \\ 2 \end{bmatrix}$

Use the calculator to compute $[A]^{-1}$ * [B].

The solution matrix is displayed as $\begin{bmatrix} -5 \\ -8 \\ 20 \end{bmatrix}$, which translates to the ordered triple(- 5,- 8, 20).

◆

What happens when the entries in the inverse of the coefficient matrix are decimal approximations? The determinant of the matrix illustrates the response.

The determinant is the numerical value assigned to a square matrix. More specifically, the determinant of a given matrix A is the denominator value for all the entries of the inverse of matrix A.

Suppose A is the 2 x 2 matrix $\begin{bmatrix} 4 & 6 \\ 8 & -9 \end{bmatrix}$.

Use the calculator to compute $[A]^{-1}$. Your display should agree with the one below. You can use the cursor arrows to cursor through each decimal entry.

$$\begin{bmatrix} .1071428571 & .0714285714 \\ .0952380952 & -.0476190476 \end{bmatrix}$$

Now compute the determinant of A. Press **[2nd] <MATRIX>**, (cursor right to highlight **MATH), [1](1:det( ) [2nd] <MATRIX> [1:](1 [A]) [ENTER] [ )]**. The determinant of A is displayed as - 84. The - 84 is the denominator of the decimal approximations displayed in $[A]^{-1}$.

TI-86	TO COMPUTE THE DETERMINANT OF A, PRESS **[2ND] <MATRX> [F3](MATH) [F1](det) [2ND] <M1>(NAMES) [F1](A) [ENTER].**

Multiplying each entry in the inverse matrix by the determinant value will yield the corresponding numerator values. Press **[2nd] <MATRIX>** (highlight **MATH**) **[1] (1:det) [2nd] <MATRIX> [1](1: [A]) [*] [2nd] <MATRX> [1](1: [A]) [x^{-1}] [ENTER].** Your screen should correspond to the one at the right.

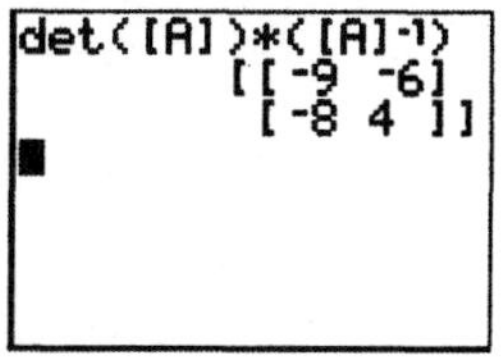

Since this matrix is the result of -84 · $[A]^{-1}$, we can conclude that the decimal approximations in $[A]^{-1}$ are represented by the following fractions:

$$\begin{bmatrix} \frac{-9}{-84} & \frac{-6}{-84} \\ \frac{-8}{-84} & \frac{4}{-84} \end{bmatrix} = \begin{bmatrix} \frac{3}{28} & \frac{1}{14} \\ \frac{2}{21} & -\frac{1}{21} \end{bmatrix}$$

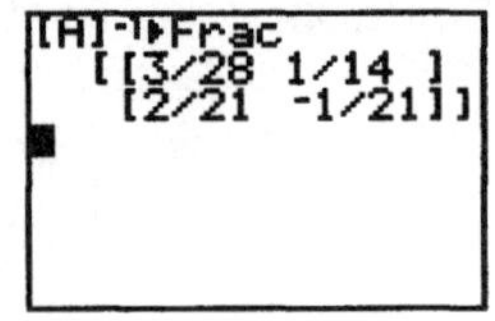

Confirm this with the calculator by computing $[A]^{-1}$ again, but this time use the ▸**FRAC** command. Your display should correspond to the one at the right.

EXERCISE SET CONTINUED

Directions: Rewrite each system in matrix multiplication form and solve by using the inverse of the coefficient matrix. Express answers in fractional form.

7. $$\begin{aligned} 3x-2y&=7 \\ 5x+2y&=3 \end{aligned}$$

$$\begin{bmatrix} \_\_ & \_\_ \\ \_\_ & \_\_ \end{bmatrix} \cdot \begin{bmatrix} \_\_ \\ \_\_ \end{bmatrix} = \begin{bmatrix} \_\_ \\ \_\_ \end{bmatrix}$$ Solution matrix: $\begin{bmatrix} \_\_ \\ \_\_ \end{bmatrix}$

Ordered pair (_____ , _____)

8. $$\begin{aligned} 8x+2y&=7 \\ 3x+12y&=5 \end{aligned}$$

$$\begin{bmatrix} \_\_ & \_\_ \\ \_\_ & \_\_ \end{bmatrix} \cdot \begin{bmatrix} \_\_ \\ \_\_ \end{bmatrix} = \begin{bmatrix} \_\_ \\ \_\_ \end{bmatrix}$$ Solution matrix: $\begin{bmatrix} \_\_ \\ \_\_ \end{bmatrix}$

Ordered pair (_____ , _____)

9. $$\begin{aligned} 4x+2y+6z&=2 \\ -7x-3y+3z&=2 \\ 3x+6y+9z&=6 \end{aligned}$$

$$\begin{bmatrix} \_\_ & \_\_ & \_\_ \\ \_\_ & \_\_ & \_\_ \\ \_\_ & \_\_ & \_\_ \end{bmatrix} \cdot \begin{bmatrix} \_\_ \\ \_\_ \\ \_\_ \end{bmatrix} = \begin{bmatrix} \_\_ \\ \_\_ \\ \_\_ \end{bmatrix}$$ Solution matrix: $\begin{bmatrix} \_\_ \\ \_\_ \\ \_\_ \end{bmatrix}$

Ordered triple: (______ , ______ , ______)

10. $\begin{array}{l} 3x+2y-z=1 \\ 2x-y+3z=5 \\ x+3y+2z=2 \end{array}$

$$\begin{bmatrix} \_\_ & \_\_ & \_\_ \\ \_\_ & \_\_ & \_\_ \\ \_\_ & \_\_ & \_\_ \end{bmatrix} \cdot \begin{bmatrix} \_\_ \\ \_\_ \\ \_\_ \end{bmatrix} = \begin{bmatrix} \_\_ \\ \_\_ \\ \_\_ \end{bmatrix}$$

Solution matrix: $\begin{bmatrix} \_\_ \\ \_\_ \\ \_\_ \end{bmatrix}$

Ordered triple: (______, ______, ______)

11. $\begin{array}{l} x+y+4z=2 \\ 2x-y+z=1 \\ 3x-2y+3z=5 \end{array}$

$$\begin{bmatrix} \_\_ & \_\_ & \_\_ \\ \_\_ & \_\_ & \_\_ \\ \_\_ & \_\_ & \_\_ \end{bmatrix} \cdot \begin{bmatrix} \_\_ \\ \_\_ \\ \_\_ \end{bmatrix} = \begin{bmatrix} \_\_ \\ \_\_ \\ \_\_ \end{bmatrix}$$

Solution matrix: $\begin{bmatrix} \_\_ \\ \_\_ \\ \_\_ \end{bmatrix}$

Ordered triple: (______, ______, ______)

✍12. What happens when the coefficient matrix is singular (i.e. has no inverse)?

$\begin{bmatrix} 1 & -4 & 1 \\ 2 & -7 & -2 \\ 3 & -11 & -1 \end{bmatrix}$ is a singular matrix. Compute the determinant. (determinant = _____)

Explain <u>why</u> a matrix with no inverse would have a zero determinant (or conversely why a zero determinant would mean the matrix is singular).

When the coefficient matrix is singular, Gaussian elimination must be used to solve the system since the coefficient matrix would have no inverse.

Example 3: Solve the system $\begin{array}{l} x-4y+z=1 \\ 2x-7y-2z=-1 \\ 3x-11y-z=2 \end{array}$ using matrices.

Solution: The matrix formed from the coefficients of the system

$\begin{array}{l} x-4y+z=1 \\ 2x-7y-2z=-1 \\ 3x-11y-z=2 \end{array}$ has no inverse. Thus, Gaussian elimination will be used to solve the system. The matrix formed by both the coefficients and the constants is an augmented matrix.

Enter the augmented matrix in the calculator as matrix B now: $\begin{bmatrix} 1 & -4 & 1 & 1 \\ 2 & -7 & -2 & -1 \\ 3 & -11 & -1 & 2 \end{bmatrix}$

NOTE: Because the rows represent individual equations: a) any row may be multiplied by a non-zero number, b) any two rows can be interchanged, and c) any two rows can be added together. Row operations are found under the **MATH** submenu of the **MATRX** menu.

TI-86	ROW OPERATIONS ARE FOUND UNDER THE **OPS** SUBMENU OF THE **MATRX** KEY.

Perform row operations on the preceeding matrix in order to place the matrix in reduced row eschelon form :

$$\begin{bmatrix} 1 & \_\_ & \_\_ & \_\_ \\ 0 & 1 & \_\_ & \_\_ \\ 0 & 0 & 1 & \_\_ \end{bmatrix}$$

a. Objective: Row 1, column 1 should have an entry of **1** with the remainder of the column being zeroes.

mathematical operation: -2 * Row 1 + Row 2 → Row 2 (i.e.replaces)

calculator operation: ***row+** (-2, [B],1,2) **STO▸ [C]**

```
*row+(-2,[B],1,2
)→[C]
 [[1  -4   1   1 ]
  [0  1   -4  -3]
  [3  -11 -1   2 ]]
```

TI-86	THE TI-86 EQUIVALENT TO ***row+** IS **mRAdd**.

NOTE: Row operations do not change the matrix stored in memory! The new matrix must be stored each time row operations are performed. The new matrix was stored in [C] to preserve the original matrix.

mathematical operation: -3 * Row 1 + Row 3 → Row 3

calculator operation: * **row+** (-3, [C],1,3) **STO▸ [C]**

```
*row+(-3,[C],1,3
)→[C]
   [[1  -4  1   1 ]
    [0  1   -4  -3]
    [0  1   -4  -1]]
```

b. Objective: Row 2, column 2 should have an entry of **1** with the remainder of the column being zeroes.

mathematical operation: -1 * Row 2 + Row 3 → Row 3

calculator operation: ***row+** (-1, [C],2,3)

```
*row+(-1,[C],2,3
)→[C]
   [[1  -4  1   1 ]
    [0  1   -4  -3]
    [0  0   0   2 ]]
```

c. The matrix yields the remaining equations:

$$\begin{aligned} x-4y+z&=1 \\ y-4z&=-3 \\ 0x+0y+0z&=-2 \end{aligned}$$

The last equation indicates that there is no solution to this system. (Had there been a solution you would need to "back" substitute at this point to find the values of x and y.)

d. The solution is the empty set.

Remember: There is a unique solution to an $n \times n$ system if and only if the coefficient matrix is nonsingular.

◆

NOTE: Not all of the available row operations were used. The following row operations indicate the order in which information is to be entered.

➻ **rowSwap** (matrix, row 1, row 2) swaps row 1 and row 2

TI-86 **rSwap** (MATRIX, ROW 1, ROW 2)

➻ **row+** (matrix, row 1, row 2) adds row 1 and row 2 and stores result in row 2

TI-86 **rAdd**(MATRIX, ROW 1, ROW 2)

➻ ***row** (value, matrix, row) multiplies a row by the indicated value

TI-86 **multR**(VALUE, MATRIX, ROW)

➻ ***row+** (value, matrix, row 1, row 2) multiplies the matrix row 1 by the indicated value, adds this product to row 2 and stores the result in row 2

TI-86 **mRAdd** (VALUE, MATRIX, ROW 1, ROW 2)

* * * * * *

The calculator can be used to write the system in both *row echelon form* and *reduced row echelon form* without individually processing the row operations. If this is acceptable to your instructor, it is the quickest way to determine the solution to the system.

Example 4: Determine the solution to the system
$$\begin{aligned}-2x+2y-3z&=4\\ x+2y-z&=-3\\ 2x-4y+z&=-7\end{aligned}$$
using the **ref** (row echelon form) feature of the calculator.

Solution: Enter the matrix $\begin{bmatrix}-2 & 2 & -3 & 4\\ 1 & 2 & -1 & -3\\ 2 & -4 & 1 & -7\end{bmatrix}$ as matrix A.

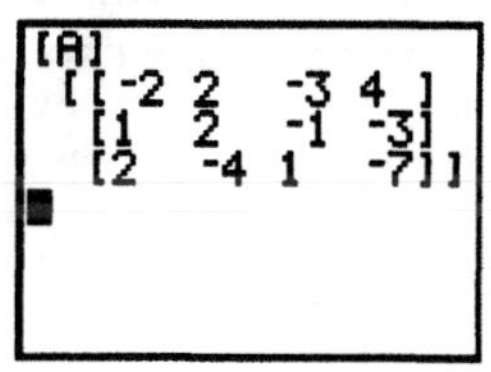

To reduce the matrix to row echelon form, **ref**, press **[2nd] <MATRIX>** , highlight **MATH** and press **[ENTER]**. Cursor down , **[▼]** , to highlight **[A](A:ref)** and press **[ENTER]**.

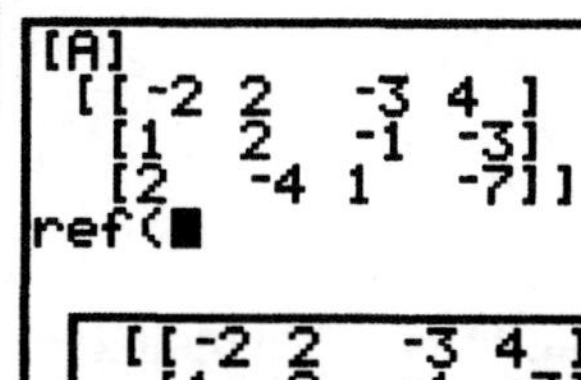

TI-86 PRESS **[2ND] <MATRIX>[F4](OPS) [F4] (REF) [( ]**

With **ref(** displayed on the home screen, enter **[A]** by pressing **[2nd]<MATRIX> [1](1:[A]) [ )] [ENTER]**.

TI-86 PRESS **[2ND] [F1](NAMES) [F1] (A) [ )]**

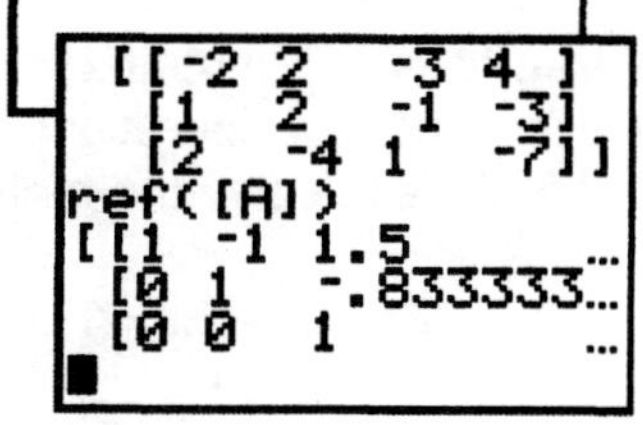

To redisplay this matrix with the entries in fractional form, press **[2nd] <ENTRY>** to retrieve the last entry, and enter the *convert to frac* command at the end of the entry. Cursor to the right to display the complete matrix.

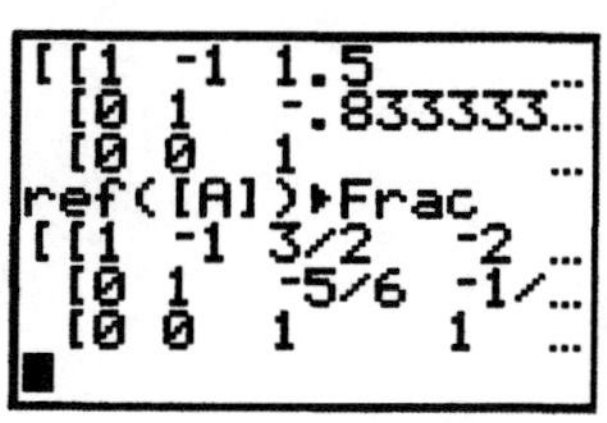

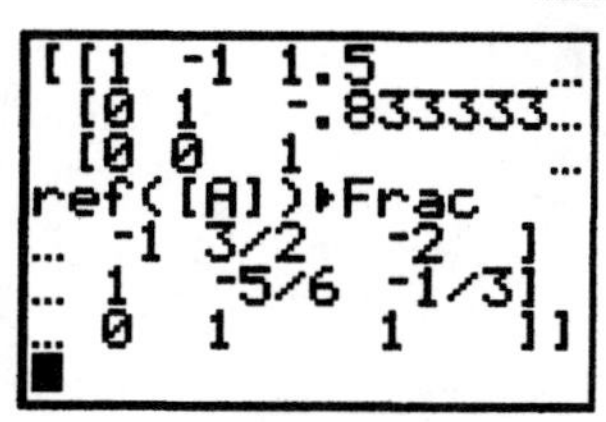

Back substitution can now be used to solve the system:

$$z = 1 \qquad \begin{aligned} y-\frac{5}{6}z&=-\frac{1}{3} \\ y-\frac{5}{6}(1)&=-\frac{1}{3} \\ y&=\frac{1}{2} \end{aligned} \qquad \begin{aligned} x-y+\frac{3}{2}z&=-2 \\ x-\frac{1}{2}+\frac{3}{2}(1)&=-2 \\ x&=-3 \end{aligned}$$

The solution is $\left(-3,\frac{1}{2},1\right)$. ◆

Example 5: Determine the solution to the system $\begin{aligned} -2x+2y-3z&=4 \\ x+2y-z&=-3 \\ 2x-4y+z&=-7 \end{aligned}$ using the **rref** (reduced row echelon form) feature of the calculator.

Solution: Enter the matrix $\begin{bmatrix} -2 & 2 & -3 & 4 \\ 1 & 2 & -1 & -3 \\ 2 & -4 & 1 & -7 \end{bmatrix}$ as matrix A.

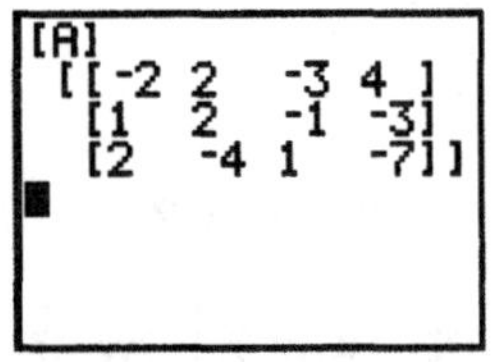

To put the matrix in reduced row echelon form, **rref**, press **[2nd] <MATRIX>**, highlight **MATH** and press **[ENTER]**. Cursor down , **[▼]**, to highlight **[B](B:rref)** and press **[ENTER]**.

TI-86 PRESS **[2ND] <MATRIX>[F4](OPS) [F5] (RREF) [( ]**

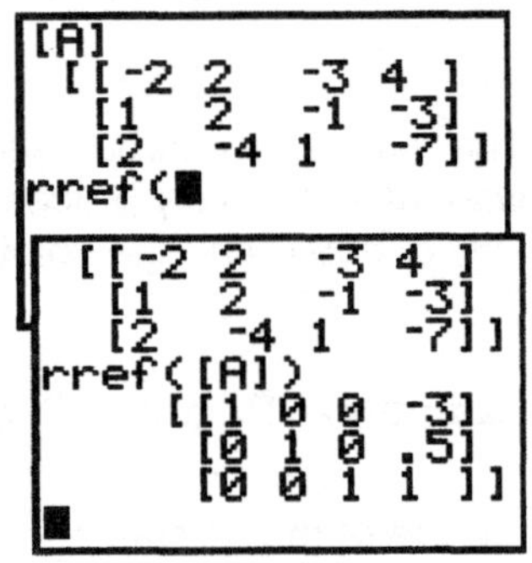

With **rref(** displayed on the home screen, enter **[A]** by pressing **[2nd]<MATRIX> [1](1:[A]) [ )] [ENTER]**.

TI-86 PRESS **[2ND] [F1](NAMES) [F1] (A) [ )]**

In reduced row echelon form, the solutions can be read from the matrix, x = -3, y = 0.5 and z = 1: (-3, 0.5, 1)

◆

EXERCISE SET CONTINUED

Directions: Use matrix row operation to perform Gaussian elimination to solve the following systems. Record the row operations as was done in Example 2. Verify your row operations by using either the **rref** or **ref** feature to solve the matrix directly.

13. $\begin{aligned} 2x+y+2z&=1 \\ x-2y+3z&=4 \\ 2x-3y+z&=0 \end{aligned}$

Solution: ________________

14. $\begin{aligned} 4x - 4y - 3z &= 0 \\ 4x + 3y - 3z &= 0 \\ 4x + 6y - 3z &= 1 \end{aligned}$

Solution: ____________________

15. $\begin{aligned} x + y + 9z &= 8 \\ x + 3y - z &= 0 \\ x + 6y - 7z &= 0 \end{aligned}$

Solution: ____________________

16. $\begin{aligned} 2x + 3y - z &= -2 \\ x + 2y + 2z &= 8 \\ 5x + 9y + 5z &= 2 \end{aligned}$

Solution: ____________________

17. **Application:** The atomic number lead is four more than three times the atomic number of iron. If the atomic number of lead is decreased by twice the atomic number of iron the result is the atomic number of zinc which is 30. Find the atomic numbers of lead and iron.

APPLICATION TO COMPLEX NUMBERS

The set of all matrices of the form $\begin{bmatrix} a & b \\ -b & a \end{bmatrix}$ with the usual operations of addition and

multiplication of matrices and the set of all complex numbers a + bi with their operations of addition and multiplication are algebraically equivalent (i.e. isomorphic). If the complex number a + bi is identified with the matrix $\begin{bmatrix} a & b \\ -b & a \end{bmatrix}$ and the complex number c + di with the matrix $\begin{bmatrix} c & d \\ -d & c \end{bmatrix}$ then the matrix sum $\begin{bmatrix} a & b \\ -b & a \end{bmatrix} + \begin{bmatrix} c & d \\ -d & c \end{bmatrix} = \begin{bmatrix} a+c & b+d \\ -(b+d) & a+c \end{bmatrix}$ is identified with the complex sum (a + bi) + (c + di). This complex sum is equal to (a + c) +(b + d)i. Thus the complex product (3 - 2i)(4 + 5i) could be performed as the matrix multiplication of

$\begin{bmatrix} 3 & -2 \\ 2 & 3 \end{bmatrix} \bullet \begin{bmatrix} 4 & 5 \\ -5 & 4 \end{bmatrix}$. This multiplication yields $\begin{bmatrix} 2 & 7 \\ -7 & 2 \end{bmatrix}$. Thus the product of the complex numbers is 2 + 7i.

EXERCISE SET CONTINUED

Directions: Use matrix operations to perform the following complex number computations. Express each complex number as a matrix and display the matrix (with fractional entries where applicable) and the corresponding complex number that result from the matrix computation.

18. (4 - 3i) + (7 - 2i)

19. (5 + 6i) - (-7 - 3i)

20. (5 - 7i) (3 - 4i)

21. Find the reciprocal (multiplicative inverse) of 1 + 2i.

22. $\dfrac{4+7i}{3-2i}$ (HINT: Use multiplicative inverses.)

Solutions: **1.** no **2.** Although the product may be defined, number of columns in matrix A equals number of rows in matrix B, AB does not always equal to BA.

3. Both matrices should be $\begin{bmatrix} 65 & 37 \\ 33 & 70 \end{bmatrix}$. Matrix multiplication is distributive from the right.

4. Both matrices should be $\begin{bmatrix} 7 & 57 & 73 \\ -9 & 43 & 13 \\ 17 & 38 & 85 \end{bmatrix}$. Matrix multiplication is distributive from the left.

5. Both matrices should yield a dimension error . A matrix is distributive through multiplication provided the dimension requirements are satisfied. However, A(B+C) does not always equal to (B +C)A.

6. "Beagles": $14220, "Herman's Hideaways": $10,245, "Peter, Piper & Pepper": $8875

7. (1,-2) **8.** (37/45, 19/90) **9.** (-7/13, 6/13, 7/13) **10.** (6/7, -2/7, 1)

11. (-3/2, -5/2, 3/2) **12.** determinant = 0 **13.** (-1, -1/7, 11/7) **14.** ∅; Answers may vary.

15. (-20/3, 8/3, 4/3) **16.** ∅ **17.** Lead = 82, Iron = 26

18. 11 - 5i **19.** 12 + 9i **20.** -13 - 41i **21.** $(1+2i)^{-1}$ = 1/5 - (2/5)i

22. -2/13 + (29/13)i NOTE: $\frac{A}{B} = A \div B = A \bullet \frac{1}{B} = A \bullet B^{-1}$

UNIT 23
TRIGONOMETRIC FUNCTIONS:
AMPLITUDE, PHASE SHIFTS, AND TRANSLATIONS

*Prerequisite: Unit #17

Review

When translations and stretches of graphs were previously examined, the following conclusions were drawn:

a. $y = f(x) + c$: translates vertically c units from the graph of $y = f(x)$, possibly affecting the range
b. $y = f(x + b)$: translates horizontally b units from the graph of $y = f(x)$, possibly affecting the domain
c. $y = a \cdot f(x)$: compresses or stretches the graph, reflects the graph over the x-axis

Graphs Involving Sine and Cosine

The sine and cosine functions are periodic functions. As the value of x increases, the values of the two functions continuously repeat. When graphing the functions, the period and amplitude of the functions are the focal point. The period of the function is the value p where $f(x + p) = f(x)$, and the amplitude is the maximum distance of the curve from the axis (determined by (maxY - minY)/2). The calculator MODE can either be set to **Radian** or **Degree.** In radian mode, the x-values will be rounded values of a constant times π. Thus when determining the period of a function by examining the graph display, having the calculator set to **Degree** MODE facilitates the process. Before proceeding, set **MODE** to **Degree** and the viewing WINDOW to **ZTrig.** In **ZTrig**, the Xscl = 90 and the min and max values are such that each move of the cursor is equal to 7.5 degrees. It is important to remember that **ZTrig** must be reset each time the calculator is changed from **Degree** to **Radian.**

✍1. Use TRACE to establish the period and amplitude of $y = \sin x$ and $y = \cos x$.

$y = \sin x$

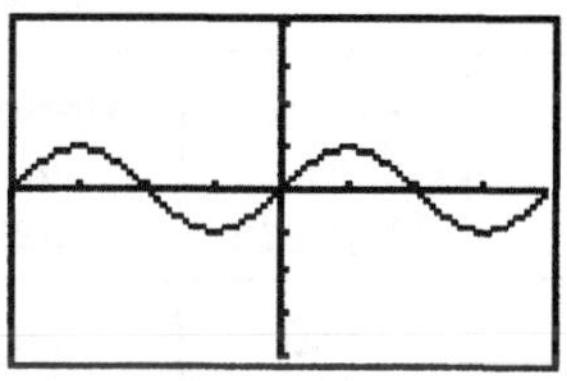

amplitude:_______________

period:_________________

symmetric with respect to:__________

$y = \cos x$

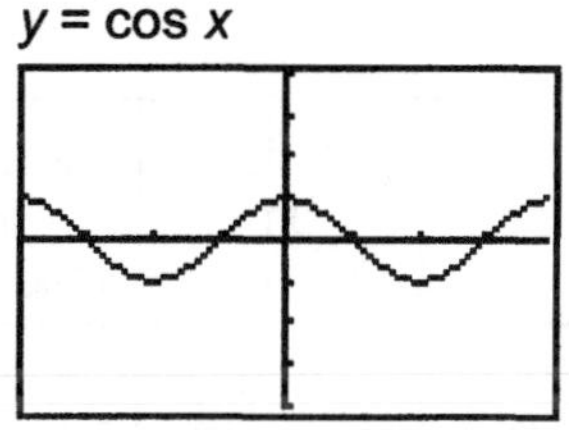

amplitude:_______________

period:___________________

symmetric with respect to:_________

What is the basic difference in the graphs?

2. Each of the groups of functions below is in the form $y = a \cdot f(x)$. Sketch each graph in a different color on the display below. The reference graph, listed first in the series, is graphed for you.

$y = \sin x$, $y = 3 \sin x$,
$y = 0.5 \sin x$

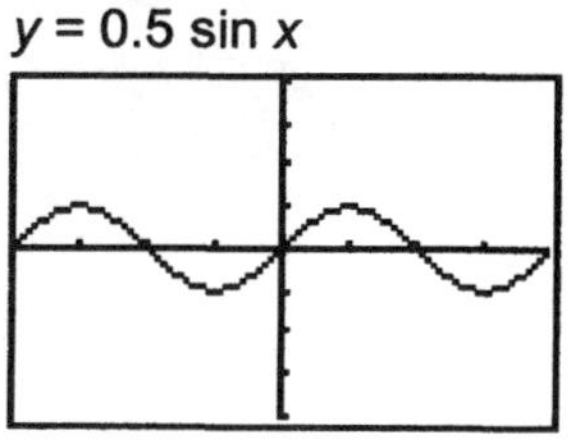

$y = \cos x$, $y= 3 \cos x$,
$y = 0.5 \cos x$

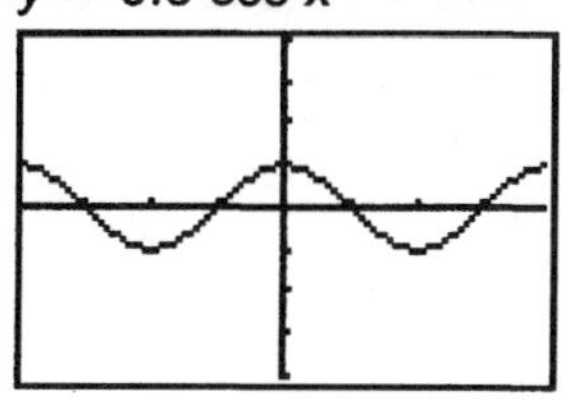

✍ 3. In general, the amplitude of $y = a \cdot f(x)$ is determined by _______.

If $y = \sin x$ and $y = \cos x$ have extreme values of y such that $-1 \le \sin x \le 1$ and $-1 < \cos x \le 1$, respectively, then $y = a{\cdot}\sin x$ and $y = a{\cdot}\cos x$ have extreme values of y such that ____________________ and ____________________, respectively.

Does $a > 0$ affect the period of either curve?

Discuss the relationship between the amplitude of the function and the range of the function.

✍ 4. Compare the graphs of $y = 3 \sin x$ and $y = -3 \sin x$. Are either the period or amplitude affected when $a < 0$?

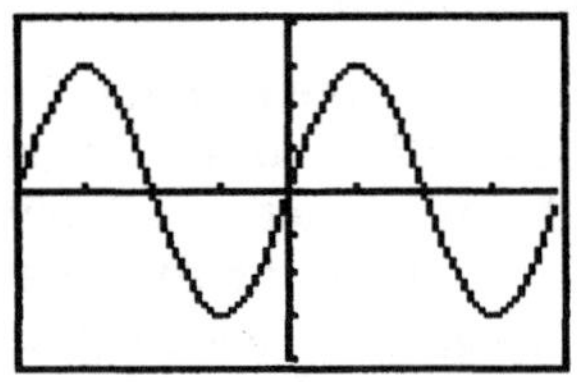

$y = 3 \sin x$

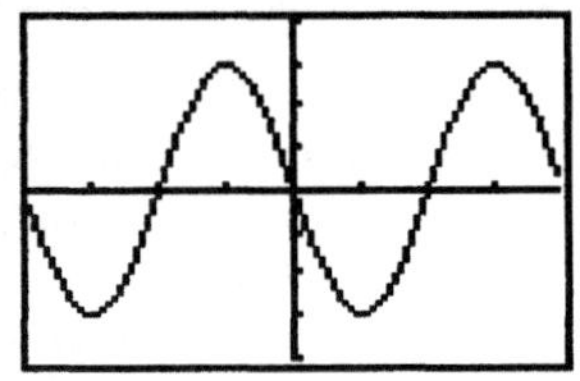

$y = -3 \sin x$

In general, the graph is reflected across the x-axis when $a < 0$.

✍ 5. Individually graph each of the functions that is in the form $y = f(bx)$. Use TRACE to examine the effect of $b > 0$ on the period of the function. Record your findings both in terms of degrees and radians (π format), in the following table.

Function	Period in Degrees	Period in Radians
$y = \sin x$	360	2π
$y = \sin .75x$		
$y = \sin .5x$		
$y = \sin .25x$		
$y = \sin 2x$		
$y = \sin 3x$		
$y = \sin 4x$		

Function	Period in Degrees	Period in Radians
$y = \cos x$	360	2π
$y = \cos .75x$		
$y = \cos .5x$		
$y = \cos .25x$		
$y = \cos 2x$		
$y = \cos 3x$		
$y = \cos 4x$		

State the algebraic relationship between radians and degrees that is used in converting from one method of measurement to the other.

✍ 6. Compare the graphs of $y = \sin 2x$, $y = \sin -2x$ and $y = \cos 2x$, $y = \cos -2x$. Sketch each graph in a different color on the display below. The first graph in each pair is already displayed.

$y = \sin 2x$, $y = \sin -2x$

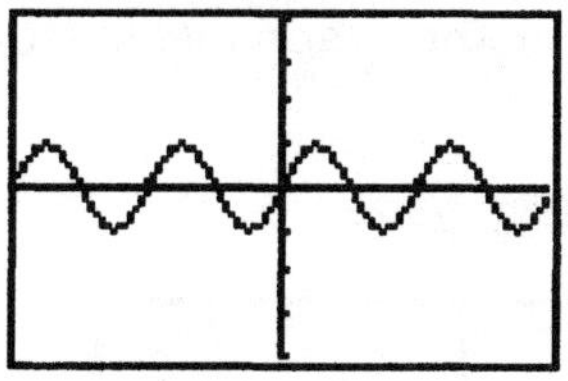

$y = \cos 2x$, $y = \cos -2x$

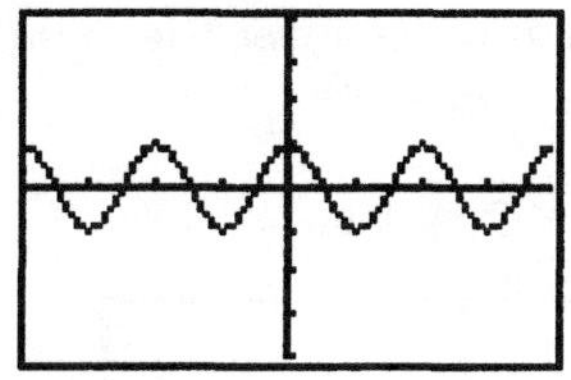

Explain the effect (or lack of effect) of $b < 0$ on each pair of graphs.

Classify the sine and cosine functions as either even or odd. Will the size of b affect the classification of even or odd?

✍ 7. Each group of functions below is in the form $y = (x + c)$. Sketch each graph in a different color on the display below. The reference graph is listed first in the series and is already displayed.

$y = \sin x$, $y = \sin (x + 45)$,
$y = \sin (x - 45)$

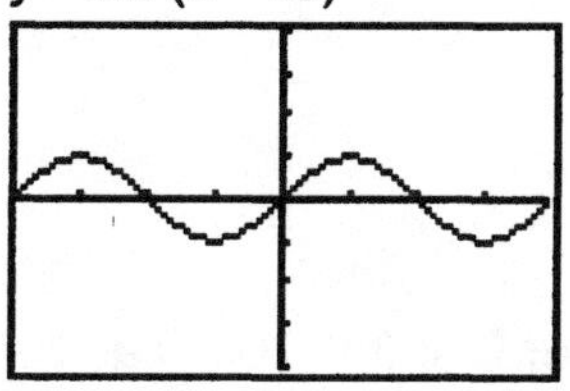

$y = \cos x$, $y = \cos (x + 45)$,
$y = \cos (x - 45)$

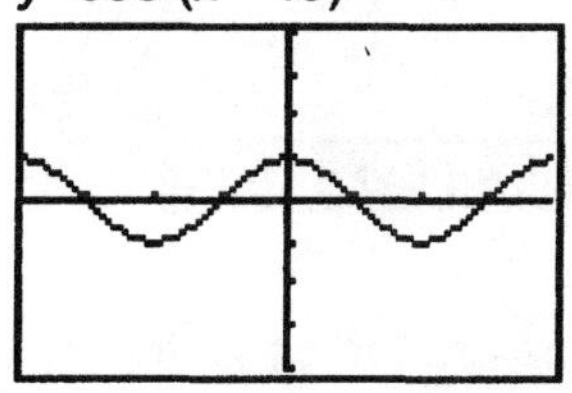

TRACE and describe the effect of c on the graph.

8. Change the **MODE** to **Radian**, reset the WINDOW to **ZTrig**, and graph the following:
$y = \sin (x + \pi/6)$, $y = \sin (x + \pi/4)$, $y = \sin (x + \pi/2)$, $y = \sin (x + \pi)$
(Suggestion: sketch each graph in a different color.)

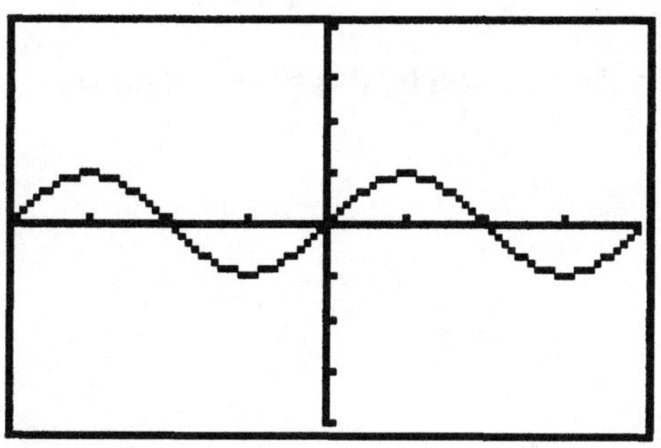

Observe the effect of the constant on the graph of the sine curve.

The value c is the phase shift of the graph. Shifting occurs based on the sign of c. When $c > 0$, the graph shifts c units to the ____________(left/right) and when $c < 0$, the graph shifts c units to the ______________(left/right).

✍ 9. Describe the effect if the constant c is a multiple of 2π.

NOTE: Change the **MODE** back to **Degree** and reset the WINDOW to **ZTrig**.

✍ 10.Each group of functions below is in the form $y = f(x) + d$. Sketch each graph in a different color on the display below. The reference graph is listed first in the series and appears on the graph.

NOTE: Be sure to enter these equations as $y = \sin(x) + 2$, etc. or the graph displayed will actually be $y = \sin(x+2)$.

$y = \sin x + 2$, $y = \sin x - 2$

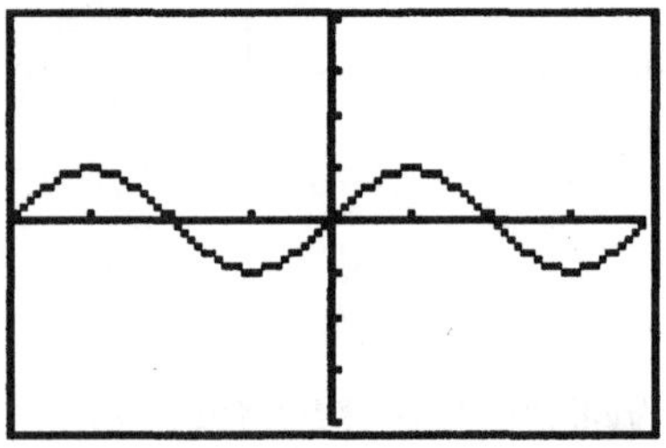

$y = \cos x + 3$, $y = \cos x - 3$

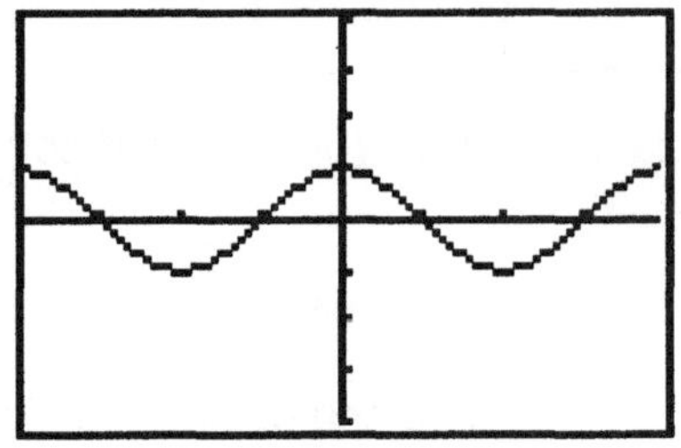

Does the constant *d* affect the period or the amplitude?

Graphs Involving Tangent and Cotangent

The sine and cosine functions are the basic trigonometric functions. The other four functions can be defined in terms of these two functions. Consider the definition of the six trigonometric functions for the acute angle x:

$\sin x = \dfrac{opposite}{hypotenuse}$ $\quad \tan x = \dfrac{opposite}{adjacent}$ $\quad \csc x = \dfrac{hypotenuse}{opposite}$

$\cos x = \dfrac{adjacent}{hypotenuse}$ $\quad \cot x = \dfrac{adjacent}{opposite}$ $\quad \sec x = \dfrac{hypotenuse}{adjacent}$

From these six functions of the acute angle X, the following are determined by the sine and cosine functions:

$\tan x = \dfrac{\sin X}{\cos X}$ $\quad \cot x = \dfrac{\cos X}{\sin X} = \dfrac{1}{\tan X}$

$\sec x = \dfrac{1}{\cos X}$ $\quad \csc x = \dfrac{1}{\sin X}$

The graphing calculator has sine, cosine and tangent function keys, however all other functions will need to be entered in terms of the above definitions. Thus $y = \cot x$ will be entered as $y = 1/\tan x$.

✍ 11. Establish the period of $y = \tan x$ and $y = \cot x$. The graph of each is displayed below. Use TRACE to determine the period.

$y = \tan x$

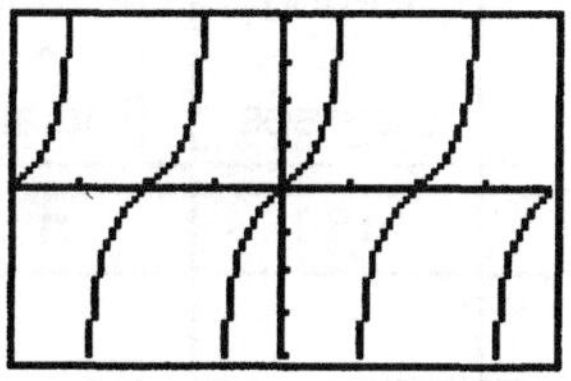

period:____________________

$y = \cot x$

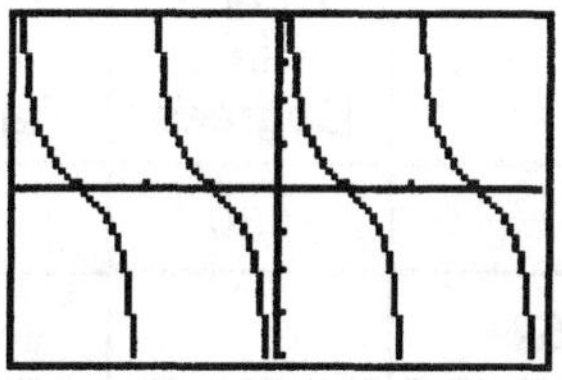

period:____________________

Describe the basic difference(s) in the graphs. Use the definitions of tangent and cotangent from the previous page to support your description.

Where do the vertical asymptotes occur? (Be aware that it may sometimes be necessary to display these graphs in dot mode.)

12. Each of the groups of functions below is in the form $y = a \cdot f(x)$. Sketch each graph in a different color on the display below. The reference graph, listed first in the series, is graphed for you.

$y = \tan x$, $y = 2 \tan x$, $y = .2 \tan x$

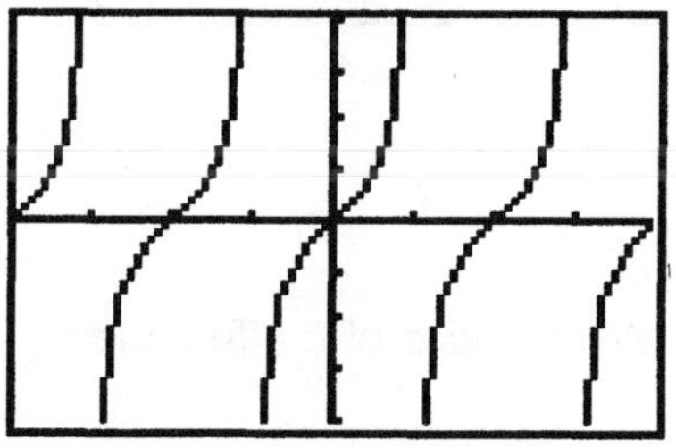

$y = \cot x$, $y = 2 \cot x$, $y = .2 \cot x$

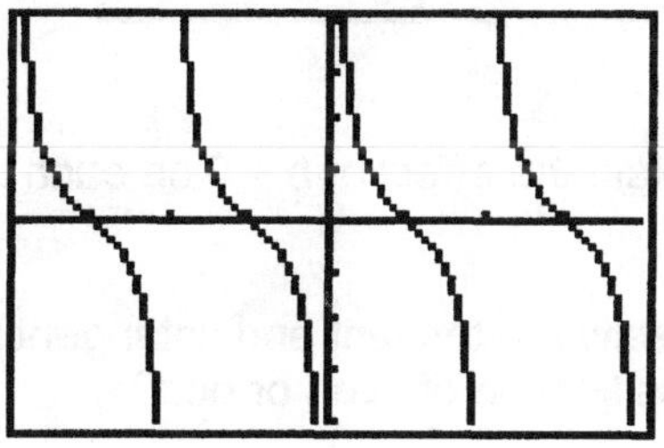

✍ 13. In general, if $a > 0$, what effect does a have on the graph of $y = a \cdot f(x)$?

✍ 14. Compare the graphs of $y = \tan x$ and $y = -\tan x$.

$y = \tan x$

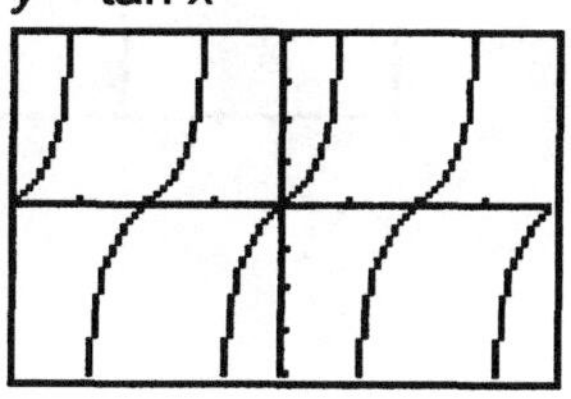

$y = -\tan x$

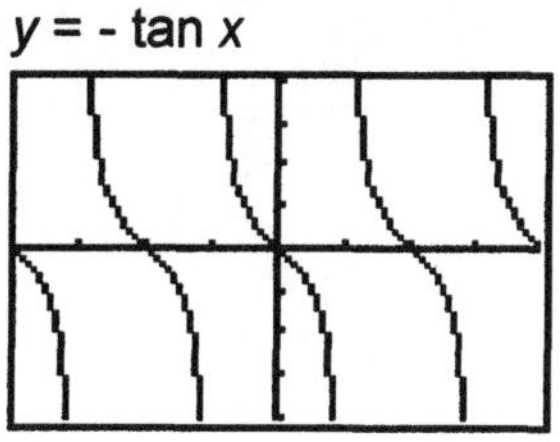

Describe the effect of $a < 0$.

15. Individually graph each of the functions that is in the form $y = f(bx)$. Use TRACE to examine the effect of $b > 0$ on the period of the function. Record your findings both in terms of degrees and radians (π format), in the following table.

Function	Period in Degrees	Period in Radians
$y = \tan x$	180	π
$y = \tan .75x$		
$y = \tan .5x$		
$y = \tan .25x$		
$y = \tan 2x$		
$y = \tan 3x$		
$y = \tan 4x$		

Function	Period in Degrees	Period in Radians
$y = \cot x$	180	π
$y = \cot .75x$		
$y = \cot .5x$		
$y = \cot .25x$		
$y = \cot 2x$		
$y = \cot 3x$		
$y = \cot 4x$		

✍16. Compare the graphs of $y = \tan 2x$, $y = \tan -2x$ and $y = \cot 2x$, $y = \cot -2x$. Sketch each graph in a different color on the display. The first graph in each pair is already displayed.

$y = \tan 2x$, $y = \tan -2x$

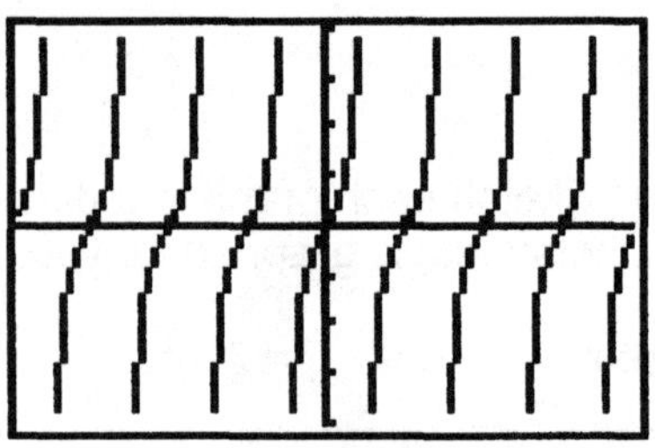

$Y = \cot 2x$, $y = \cot -2x$

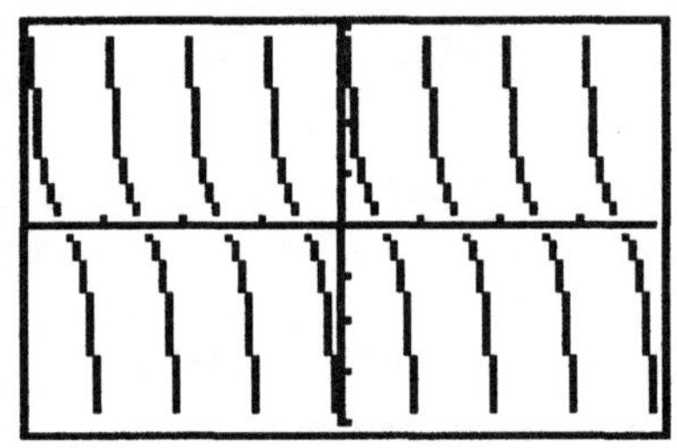

Explain the effect of $b < 0$ on each pair of graphs.

Classify the tangent and cotangent functions as even or odd. Will the size of b affect the classification of even or odd?

✍17. Graph $y = -\tan x$ and $y = \tan -x$ on the screen at the right. Justify the display.

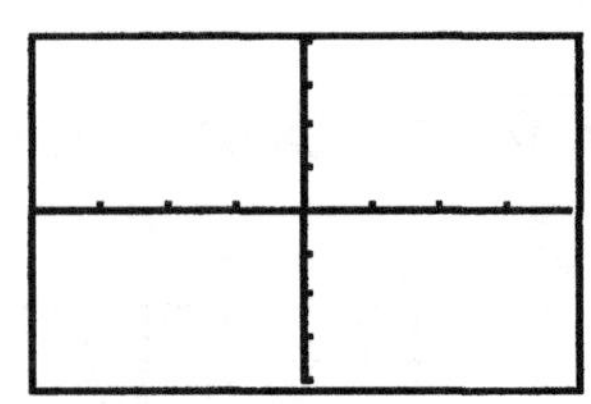

✍18. Each group of functions below is in the form $y = f(x + c)$. Sketch each graph in a different color on the display below. The reference graph is listed first in the series and graphed for you. (Dot mode will be necessary.)

$y = \tan x$, $y = \tan(x + 45)$
$y = \tan(x - 45)$

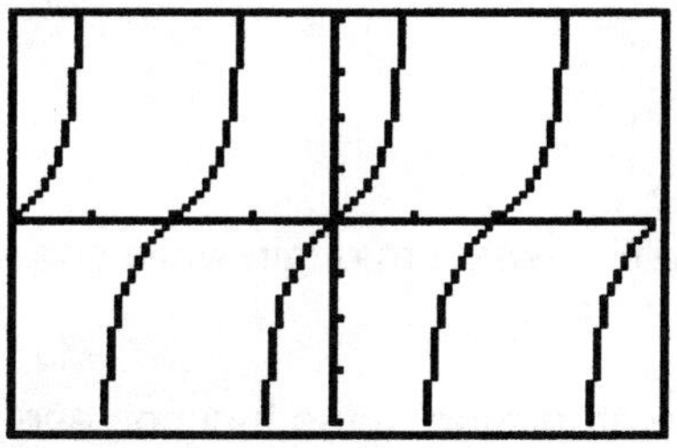

$y = \cot x$, $y = \cot(x + 45)$,
$y = \cot(x - 45)$

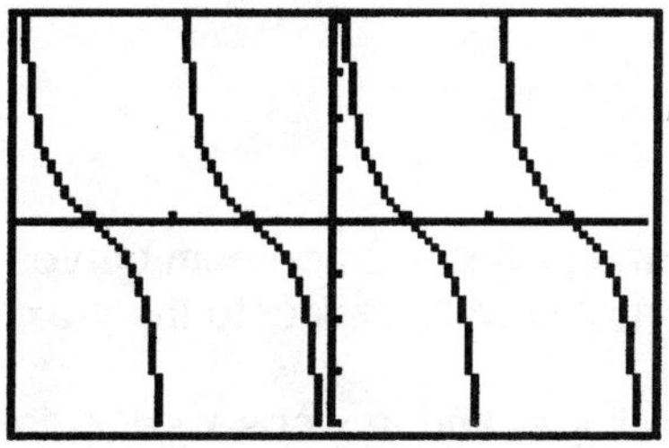

TRACE and describe the effect of c on the graph.

✍19. Explore the relationship between the graph of $y = \sin x$ and $y = \csc x$ by graphing both functions on the same viewing window. As x approaches the zeroes of $y = \sin x$, explain what happens to the graph of $y = \csc x$.

Solutions:
1. amp. = 1 and period = 2π on both curves, sine curve is symmetric with respect to origin while cosine curve is symmetric with respect to the y-axis.

3. a, -a <sin *x* ≤ a, and -a ≤ cos *x* ≤a, *a* does not affect the period; the amplitude of the function and its range are the same

4. a < 0 does not effect period or amplitude

5.

Function	Period in Degrees	Period in Radians
$y=\sin x$	360	2π
$y=\sin .75x$	480	8π/3
$y=\sin .5x$	720	4π
$y=\sin .25x$	1440	8π
$y=\sin 2x$	180	π
$y=\sin 3x$	120	2π/3
$y=\sin 4x$	90	π/2

Function	Period in Degrees	Period in Radians
$y=\cos x$	360	2π
$y=\cos .75x$	480	8π/3
$y=\cos .5x$	720	4π
$y=\cos .25x$	1440	8π
$y=\cos 2x$	180	π
$y=\cos 3x$	120	2π/3
$y=\cos 4x$	90	π/2

1 radian = 180°/π, 1 degree = π/180° radians

6. b < 0 reflects the graph across the x-axis, both functions are odd, *b* does not affect classification of even or odd

7. *c* shifts the graph left/right, c>0 shifts left whereas c<0 shifts right

8. *c* shifts the graph left/right, c>0 shifts left whereas c<0 shifts right

9. 2π shifts the graph one complete period and thus is the same graph.

10. affects the amplitude

11. both functions have period π, Vertical asymptotes appear at multiples of π/2 for the tangent function and at multiplies of π for the cotangent function.

13. The larger the value of *a*, the more the graph of the function straightens out and thus approaches the appearance of a vertical line. This occurs within each period.

14. a < 0 reflects the graph across the *y*-axis: the magnitude of |a| affects the graph the same way as in 13 above.

15.

Function	Period in Degrees	Period in Radians
$y = \tan x$	180	π
$y = \tan .75x$	240	4π/3
$y = \tan .5x$	360	2π
$y = \tan .25x$	720	4π
$y = \tan 2x$	90	π/2
$y = \tan 3x$	60	π/3
$y = \tan 4x$	45	π/4

Function	Period in Degrees	Period in Radians
$y = \cot x$	180	π
$y = \cot .75x$	240	4π/3
$y = \cot .5x$	360	2π
$y = \cot .25x$	720	4π
$y = \cot 2x$	90	π/2
$y = \cot 3x$	60	π/3
$y = \cot 4x$	45	π/4

16. $b < 0$ reflects the graph across the y-axis, The functions are both odd and the size of b does not affect this classification.

17. Answers may vary.

18. *c* shifts the graph left when added and right when subtracted

19. Answers may vary.

20. Answers may vary.

UNIT 24
GRAPHICAL EXPLORATIONS:
TRIGONOMETRIC IDENTITIES

*Prerequisite: Unit #23

Identities are equations that are true for all values of the variable (for which the expressions within the equation are defined). The graphs of the expression on each side of the equation can be used to confirm or deny the fact that a given equation is an identity.

One of the fundamental identities states that $\sin^2 x + \cos^2 x = 1$. We will verify this using the graphing calculator. First ensure all options at the far left are highlighted when **MODE** is pressed.

Example	**Graphical**	**Notation**
Verify $\sin^2 x + \cos^2 x = 1$ is an identity.	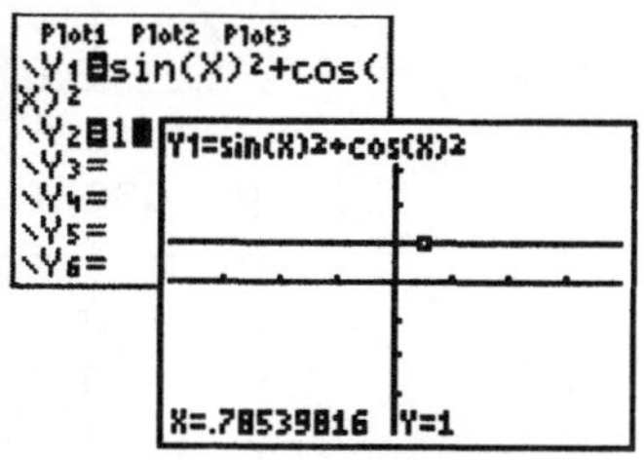 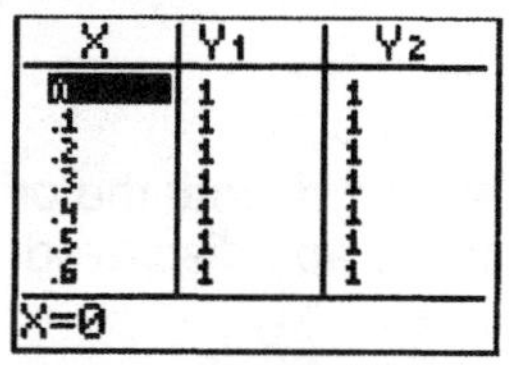	Press **[ZOOM] [7:ZTrig]** to display the graphs of the two expressions. The table confirms that the two graphs are identical since Y1 = Y2 for all values of *x*.

Enter $\sin^2 x + \cos^2 x$ at the Y1 prompt and the number 1 at the Y2 prompt. Your screen display should match the ones pictured above. To verify that this is the graph of two functions rather than one, press **[TRACE]** to locate the cursor with the coordinates displayed. Notice the equation displayed in the upper left hand corner of the screen. This indicates that the cursor is on the graph of the expression entered at the Y1= prompt. Press the down arrow key and again observe the equation in the upper left hand corner of the display screen. It is the equation entered at the Y2= prompt, however, the *x* and *y* coordinates did not change even though the cursor was moved from one graph to the other. This verifies the fact that the graph of $\sin^2 x + \cos^2 x$ is the same as the graph of the line $y = 1$ and thus the two expressions are equal. Use the table feature to verify that for each *x*-value all the values in the Y1 and Y2 columns are equal.

EXERCISE SET

✍1. Does $\sin x = \sin x + 6$? Sketch the graph on the axes provided. Explain how you obtained your answer.

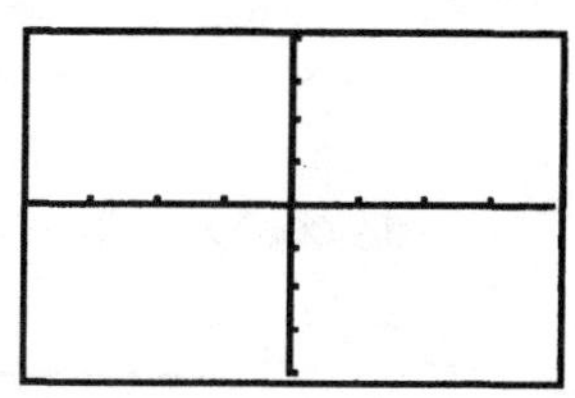

Directions: Match each of the following expressions to the appropriate graph.

_____2. $\cot x \sec x$

_____3. $\tan x \csc x$

_____4. $\dfrac{1-\cos^2 x}{\sin x}$

_____5. $\sin x \sec x$

_____6. $\dfrac{\sin x}{\cos^2 x - 1}$

_____7. $\cos x \ \csc x$

_____8. $\dfrac{\cos x}{\sin^2 x - 1}$

a. sin x

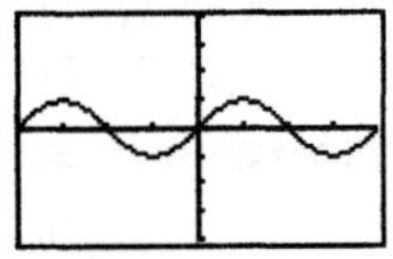

b. cot x

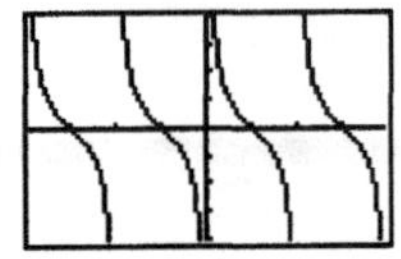

c. cos x

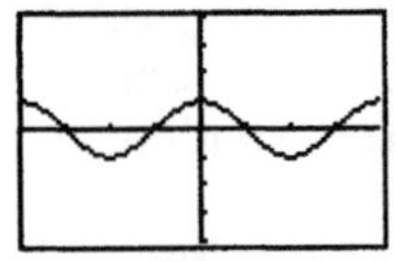

d. -sec x

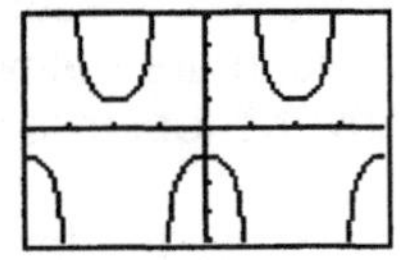

e. sec x

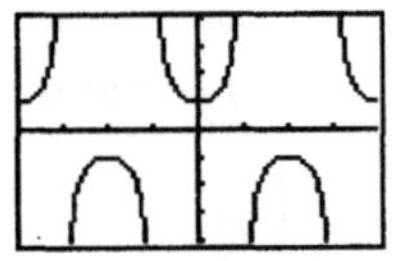

f. tan x

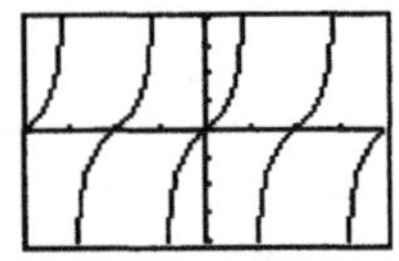

g. -csc x

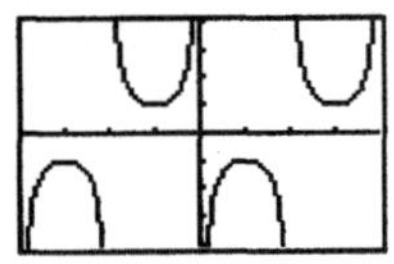

h. csc x

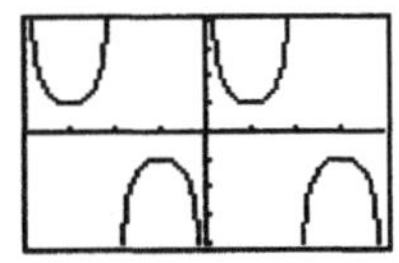

9. Factor each expression below and verify that the original problem and your factorization are graphically equivalent. Sketch the graph displayed.

a. $\sin^2 x + \sin x \cos x =$

____

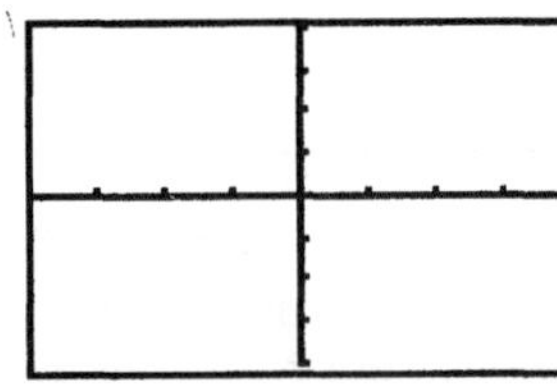

b. $\sin^2 x - 9 =$

____

c. $\sin^2 x + \sin x - 6 =$

____

d. $\cos^4 x - \sin^4 x =$

____

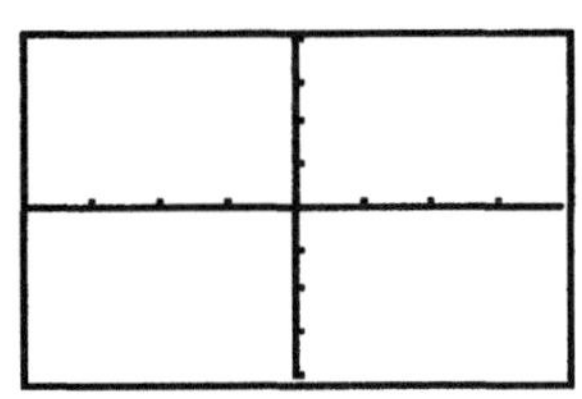

10. State the values of x that will make the equation $\cos x = \sqrt{1-\sin^2 x}$ an identity. $(0 \le x \le 2\pi)$

____

✍11. Verify with the calculator that $\cos x = \sin(\pi/2 - x)$. Record your method(s) below.

✍ 12. Is $\frac{1}{3}\tan^4 x + \frac{1}{2}\tan^2 x = \frac{1}{3}\sec^4 x - \frac{1}{6}\sec^2 x$ an identity?

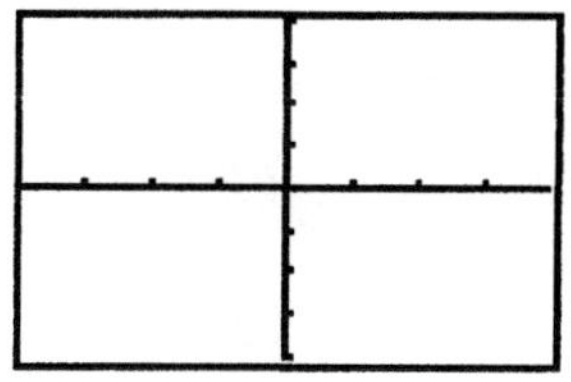

After sketching the display on the screen at the right, discuss the various methods available on the calculator for disproving the proposed identity. What constant c, expressed as a fraction, should be added to the left side of the equation to complete the identity?

Solutions: **1.** Answers may vary **2.** h **3.** e **4.** a **5.** f **6.** g **7.** b **8.** d

9. $\sin x(\sin x + \cos x)$, $(\sin x - 3)(\sin x + 3)$ window: [-4,4] by [10,1],

$(\sin x + 3)(\sin x - 2)$ window: [-6, 6] by [-10, 1], $(\cos^2 x - \sin^2 x)(\cos^2 x + \sin^2 x)$

10. $[-\pi/2, \pi/2]$ **11.** Answers may vary **12.** c = 1/6

UNIT 25
GRAPHICAL SOLUTIONS: TRIGONOMETRIC EQUATIONS

*Prerequisite: Unit #24

The focus of this unit is solving trigonometric equations. Unlike identities, which are true for all values of the variable, trigonometric equations are true for <u>some</u> values of the variable. Set the calculator to DEGREE on the MODE screen.

Example 1: Solve the equation $4 \cos x - 2 = 0$.
Algebraically: $4 \cos x = 2$ can be written as $\cos x = 1/2$

Home Screen

On the calculator, one solution of $\cos x = 1/2$ is obtained by entering $\cos^{-1}(1/2)$ on the home screen. In degree mode, $x = 60°$ or $x = 60(\pi/180) = \pi/3$ when expressed in radians. Because $\cos^{-1}(1/2)$ gives only a first quadrant solution and since cosine is also positive in quadrant IV, we also have $x = 300°$ ($11\pi/6$ radians) as an additional solution in the interval $[0,2\pi)$. This angle has 60° as its reference angle and is in quadrant IV.

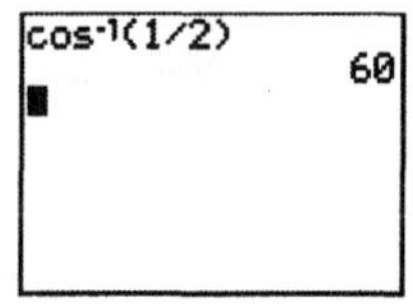

Graphically

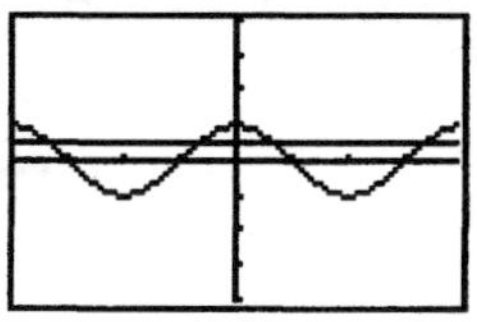

By graphing both sides of $\cos x = 1/2$, it is clear that there are infinitely many solutions. Enter $\cos x$ at the Y1 prompt and 1/2 at the Y2 prompt. Display the graph in the **ZTrig** viewing WINDOW with the MODE set to **Degree**. Using TRACE, the initial graph indicates that x has values of ±300° and ±60° (or $\pm 11\pi/6$ and $\pm\pi/3$ radians).

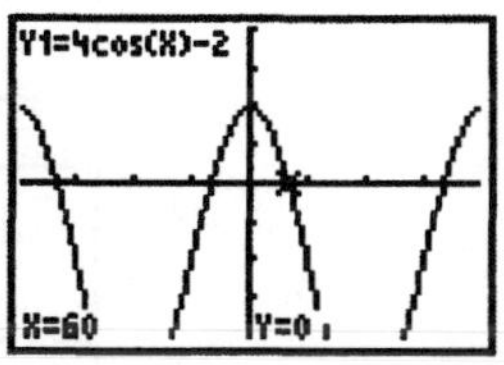

To graphically solve $4\cos x - 2 = 0$, the non-zero side is entered at the Y1 prompt and the ROOT/ZERO feature is used to calculate the solutions. Tracing confirms the previously determined values with the root of 60° ($\pi/3$) displayed on the graph.

Notation

Solutions will be displayed in the general form as $x = \pi/3 + 2n\pi$ and $x = 11\pi/6 + 2n\pi$ where n is an integer and x is expressed in radians.

Example 2

Graphically solve $\sin x = 1 - \cos 2x$ on the interval $[0,2\pi)$ using the INTERSECT option of the calculator

Graphically

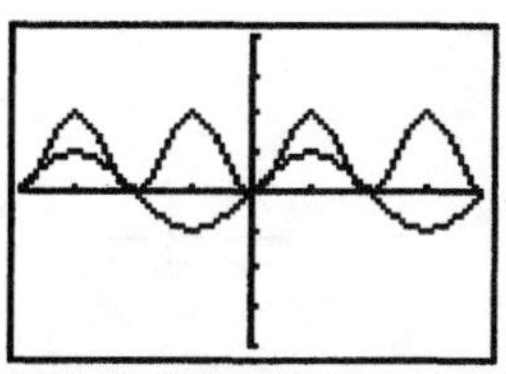

The initial graph is pictured in the **ZTrig** viewing WINDOW.

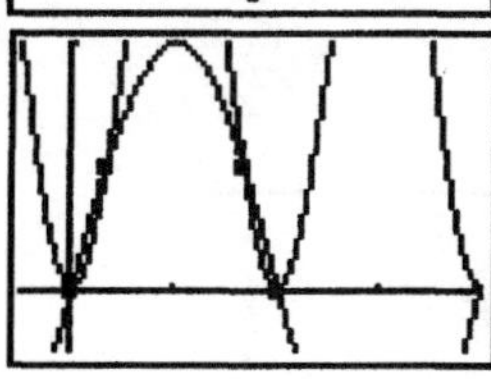

Because the points of intersection are somewhat difficult to see, a better window within the interval $[0,2\pi)$. can be determined using **ZBOX**. The solutions are 0, 30, 150 and 180 degrees.

EXERCISE SET

1. a. Algebraically solve $6\sec^2 x + 5\sec x + 1 = 0$ and confirm your algebraic solution by solving graphically.

algebraic solution:

graphical solution:

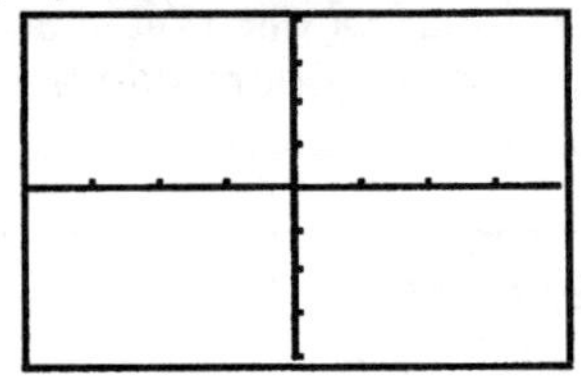

✍ b. Why are there no values of x such that $\sec x = -1/2$ or $\sec x = -1/3$?

Directions: Solve each equation graphically by the method of your choice and record the general solution in radian form. Sketch the graph displayed in the **ZTrig** viewing WINDOW (or an appropriate window that displays the solutions).

2. $4\sin^2 x = 1$

Solutions:______________________________

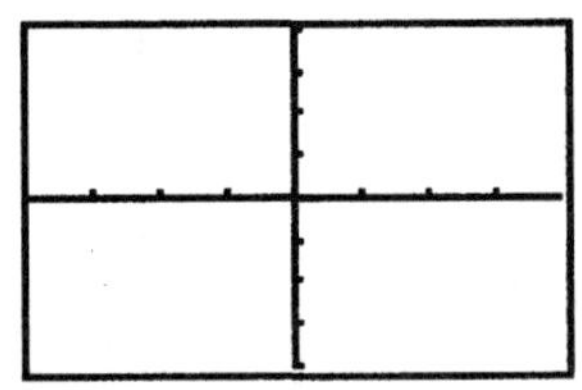

3. $2\sin x\cos x + 2\sin x + \cos x + 1 = 0$

Solutions:______________________________

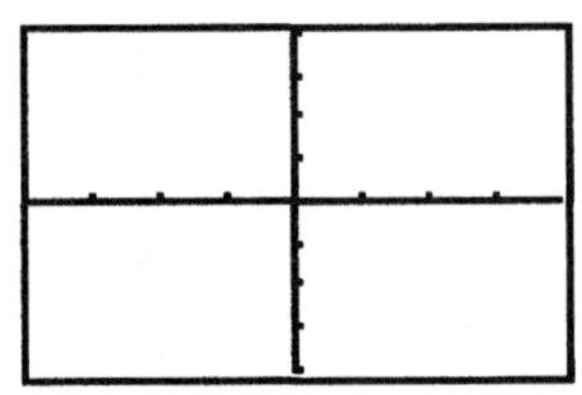

4. $7\cos 2x = \cos 4x - 8$

Solutions:______________________________

[___ , ___] by [___ , ___]

5. $2\tan x = \tan(x/2)$

Solutions:______________________________

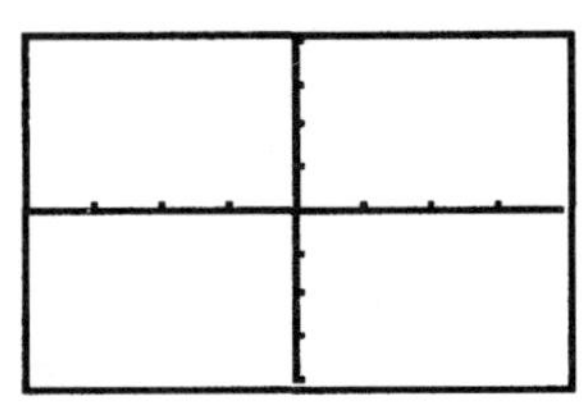

6. $\sin x + \csc x = \cos x$

Solutions:________________________________

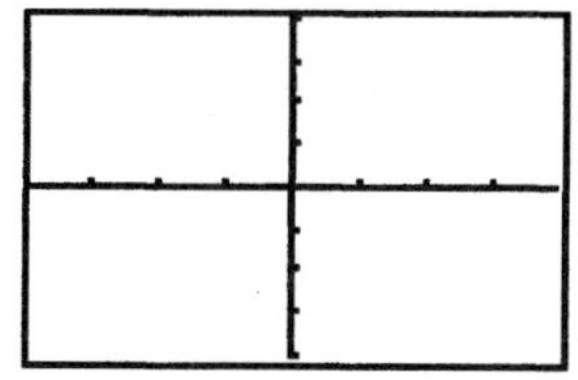

7. Find the solutions of $\sqrt{3}\sin\frac{x}{3} + \sqrt{1-\cos 2x} - \sqrt{3} = 0$, in radian measurement to the nearest hundredth in the interval [0,4π] .

Solutions:________________________________

Sketch the graph in the viewing window [-π/2,4π] by [-4,2].

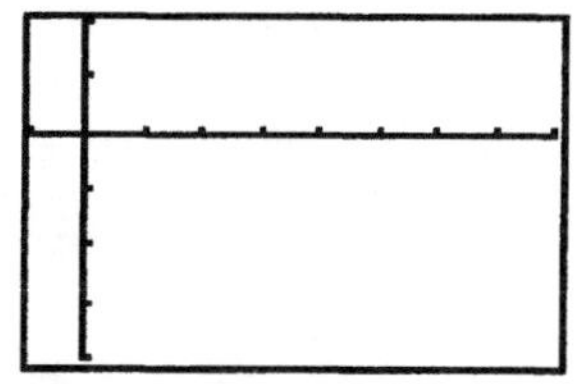

8. Examine each group of equations graphically in the **ZTrig** viewing window. Determine which ones are conditional equations and which ones represent identities. (Label ***C*** or ***I***.)

$\sin(x + 2) = \sin x + \sin 2$ _____ $\sin(x + 2) = \sin x \cos 2 + \cos x \sin 2$ _____

$\sin(x - 2) = \sin x \cos 2 - \cos x \sin 2$ _____

$\cos(x+ 2) = \cos x + \cos 2$ _____ $\cos(x + 2) = \cos x \cos 2 - \sin x \sin 2$ _____

$\cos(x- 2) = \cos x \cos 2 + \sin x \sin 2$ _____

$\tan(x + 2) = \tan x + \tan 2$ _____ $\tan(x + 2) = \dfrac{\tan x + \tan 2}{1 - \tan x \tan 2}$ _____

$\tan(x - 2) = \dfrac{\tan x - \tan 2}{1 + \tan x \tan 2}$ _____

The equations that represent identities are called the **Sum and Difference Formulas**. These identities will be useful in scientific applications.

Solutions: **1a**. ϕ **1b.** Answers may vary. **2.** $\pi/6 + 2n\pi$, $5\pi/6 + 2n\pi$, $7\pi/6 + 2n\pi$, $11\pi/6 + 2n\pi$

3. $\pi + 2n\pi$, $11\pi/6 + 2n\pi$ **4.** $\pi/2 + 2n\pi$, $3\pi/2 + 2n\pi$, WINDOW: [-6, 6] by [-10, 4]

5. $0 + 2n\pi$ **6.** ϕ

7. 0.98, 2.93, 3.28, 6.15, 6.49, 8.44

8. The conditional equations are $\sin(X+2) = \sin x + \sin 2$, $\cos(x+2) = \cos x + \cos 2$, and $\tan(x+2) = \tan x + \tan 2$. **9.** Answers may vary.

UNIT 26
POLAR GRAPHING

*Prerequisite: Unit #23

In the rectangular coordinate system, each point *P* in the coordinate plane is associated with an ordered pair of real numbers (*x*,*y*). The directed distance from the *y*-axis is represented by *x*, and *y* represents the directed distance from the *x*-axis. In polar graphing, this point *P* is associated with an ordered pair (r, θ) where *r* is the directed distance from the origin and the measure of the angle between the positive *x*-axis and the ray from the origin through point *P* is designated as θ. If θ includes all angles, then (r, θ), a coordinate of point *P*, is also represented by $(r, \theta + k \cdot 360°)$ where *k* is a positive integer.

Begin this unit by setting MODE to **Degree** and **Pol** for polar graphing. If the MODE is set for polar graphing, then pressing **[ZOOM] [6] (6:ZStandard)** will automatically set the following WINDOW values:

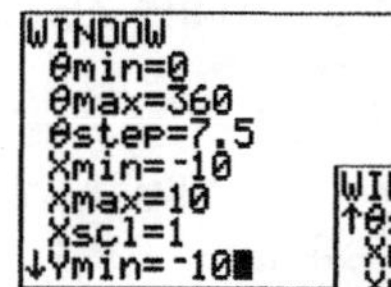

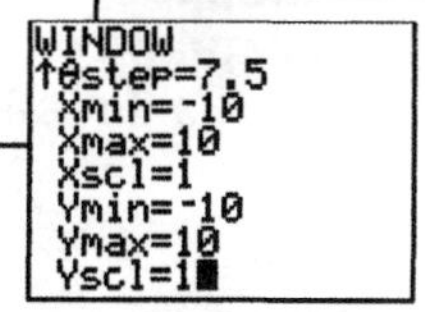

You will explore polar graphing through a series of discovery exercises.

TI-86	MODE SHOULD BE SET TO HIGHLIGHT DEGREE, **POLARC**, AND **POL**. THETA IS LISTED ON THE DISPLAYED MENU ON THE MENU BAR WHEN **[R(θ) =]** IS PRESSED.

EXERCISES

✍1. Graph $r = \theta$. Sketch the graph on the display at the right. Without adjusting the viewing window, make a conjecture about the type of graph displayed. (i.e. Is the appearance linear, parabolic, exponential, logarithmic, etc.?)

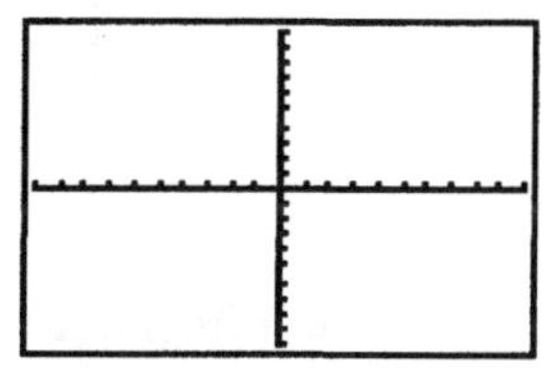

✍2. Set Xscl = 0 and Yscl = 0 and with your cursor centered at the origin, ZOOM OUT once and copy the display. What window values were affected by zooming out?

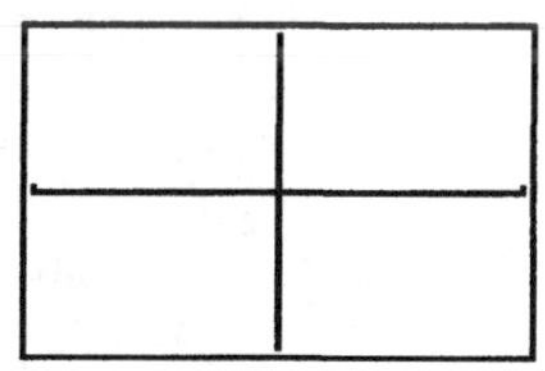

3. Again, center the cursor at the origin and ZOOM OUT. Sketch the graph display.

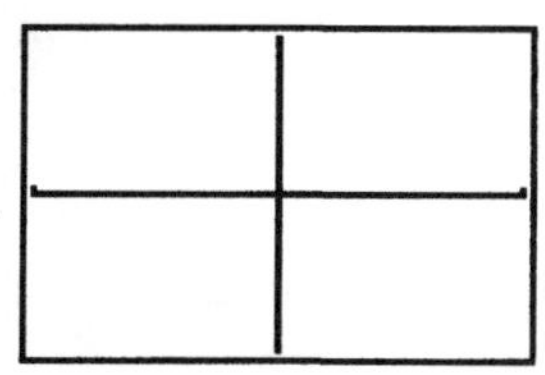

4. ZOOM OUT one final time and sketch the display. TRACE around the entire Spiral of Archimedes and observe the values of θ.

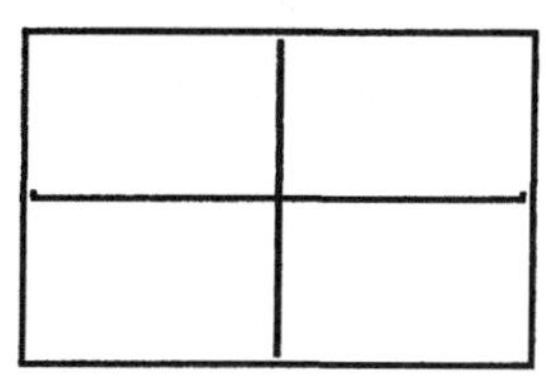

5. In the same viewing window as #4 ([-640,640] by [-640,640]), graph each of the following spirals.

$r = 2\theta$ 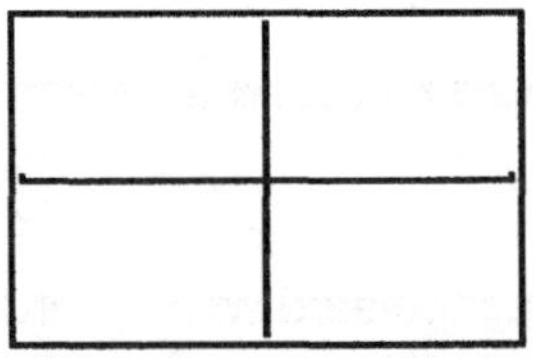

$r = -2\theta$ 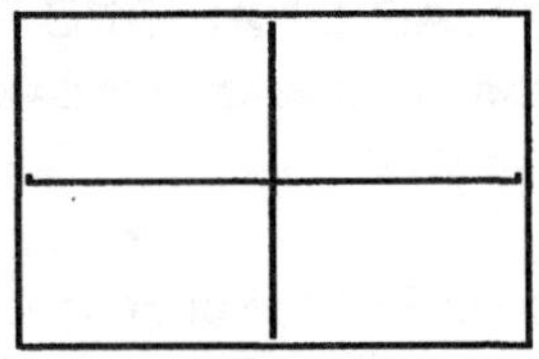

$r = \frac{1}{2}\theta$ 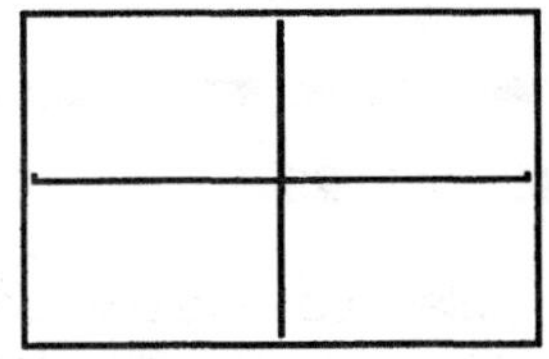

✍6. In general $r = a\,\theta$ will be a Spiral of Archimedes if $a > 0$.
The effect of a < 0 is to ______________________________.

As $|a|$ increases the spiral ______________________________.

As $|a|$ decreases the spiral ______________________________.

NOTE: Return to the ZStandard viewing window for polar graphing.

7. Graph $r = \sin 3\theta$. Set Xscl = 0 and Yscl = 0 and ZOOM IN once. Sketch the graph on the display at the right. Equations of the form $r = a \sin 3\theta$ form a three-leaved rose.

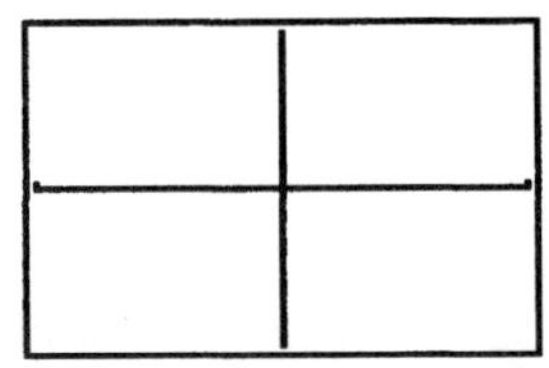

✍8. In this same viewing window, graph each of the following and sketch their display.

$r = 2 \sin 3\theta$ 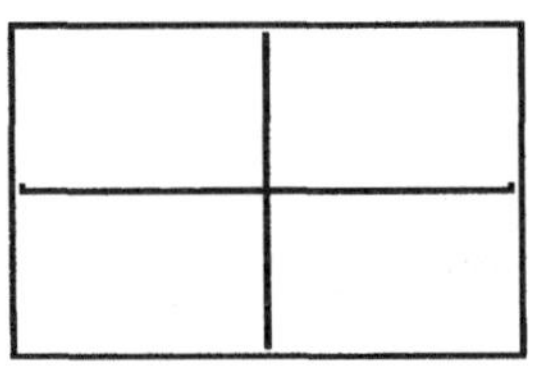

$r = -2 \sin 3\theta$ 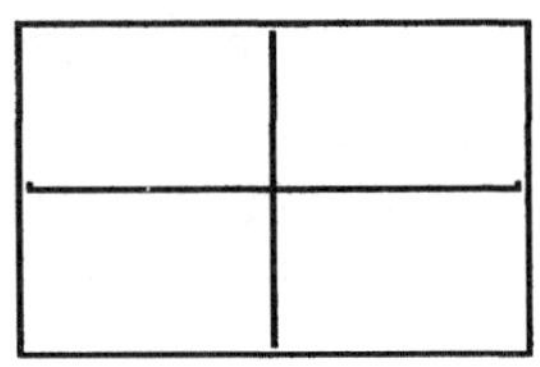

$r = \frac{1}{2} \sin 3\theta$ 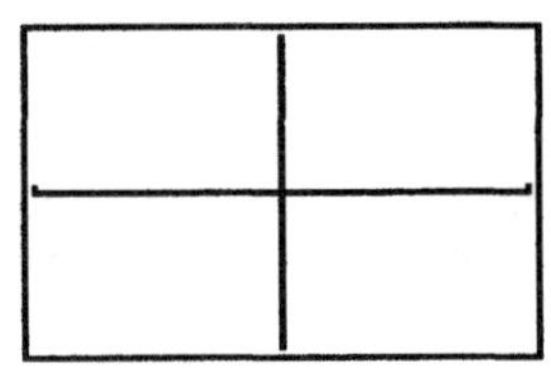

The effect of $a < 0$ is to ______________________________.

As $|a|$ increases the rose ______________________________.

As $|a|$ decreases the rose ______________________________.

✍9. Explore $r = \sin a\,\theta$ for $1 \le a \le 8$. What can you conclude about the effect of a on the graph of $a = \sin a\theta$?

10. Compare $r = \sin 2\theta$ and $r = \cos 2\theta$. Graph each on the given screen with the viewing window set to [-2.5,2.5] by [-2.5,2.5] with Xscl and Yscl = 1.

$r = \sin 2\theta$

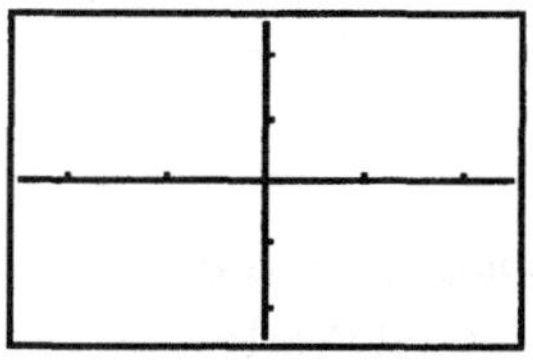

$r = \cos 2\theta$

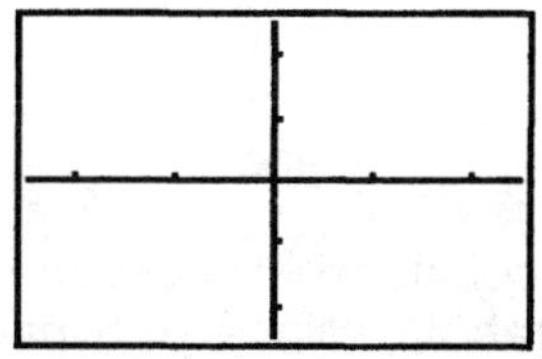

✍ Do the graphs of the polar sine and cosine have the same relationship to the *x*-axis as the sine and cosine graphs on a rectangular coordinate system have?

11. Graph $r = 1 + 2\cos\theta$ in a [-3,3] by [-3,3] polar viewing window and sketch the graph. This is the graph of a Limacon whose general form is $r = a + b\cos\theta$, $a < b$.

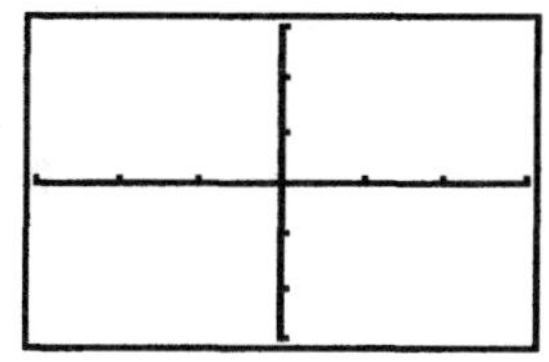

12. Graph $r = -1 + 2\cos\theta$. Is this the same graph as $r = 1 + 2\cos\theta$? (HINT: use the TABLE to verify)

✍13. Graph the following pairs of polar graphs:

$r = 2 + 3\cos\theta$
$r = 3 + 4\cos\theta$

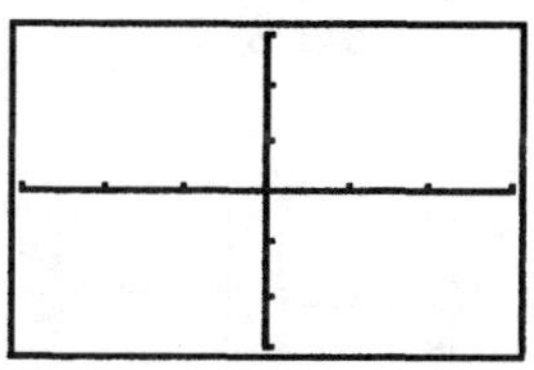

$r = 2 + 2\cos\theta$
$r = 1 + \cos\theta$

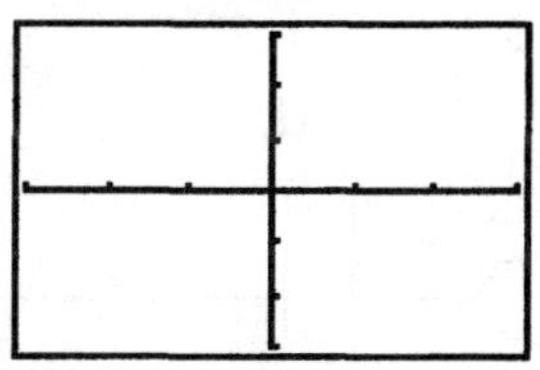

$r = 3 + 2\cos\theta$
$r = 2 + \cos\theta$

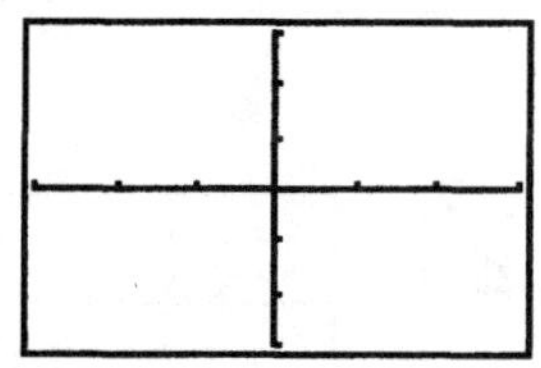

When $a < b$, as a and b both increase, the graph ______________________________.

When $a = b$, as a and b both increase, the graph ______________________________.

When $a > b$, as a and b both increase, the graph ______________________________.

Do all graphs of the form $r = a + b\cos\theta$, where $a < b$, have the same shape regardless of the size of a and b?

Do all graphs of the form $r = a + b\cos\theta$, where $a = b$, have the same shape regardless of the size of a and b?

Do all graphs of the form $r = a + b\cos\theta$, where $a > b$, have the same shape regardless of the size of a and b?

14. Graph the Hyperbolic Spiral $r = \dfrac{2}{\theta}$ using ZOOM IN or ZOOM OUT to find an appropriate viewing window.

[______ , ______] by [______ , ______]

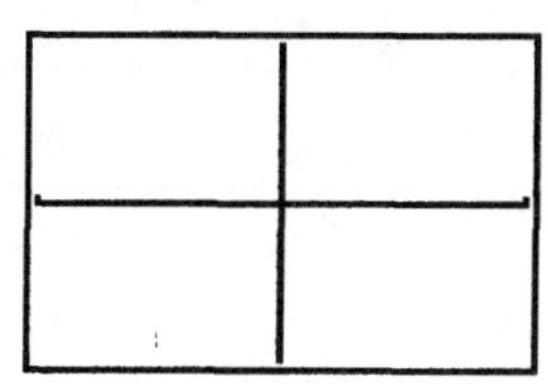

15. Graph the Logarithmic Spiral $r = 2^4\ \theta$ using ZOOM IN or ZOOM OUT to find an appropriate viewing window.

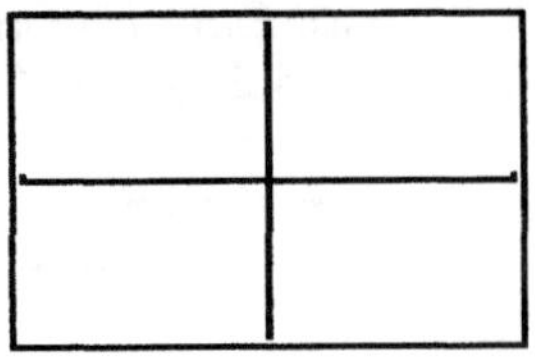

[______ , ______] by [______ , ______]

16. Polar equations in the form $r = \dfrac{ep}{1 \pm e \sin \theta}$ and $r = \dfrac{ep}{1 \pm e \cos \theta}$ represent the graphs of conics. The relationship between e and the value 1 (ie. $e < 1$, $e = 1$, or $e > 1$) determines the type of conic (parabola, ellipse, or hyperbola). Match each equation to the appropriate graph.

______i.	$r = \dfrac{4}{2 - \cos \theta}$	______vii.	$r = \dfrac{1}{1 - \cos \theta}$
______ii.	$r = \dfrac{4}{2 + \cos \theta}$	______viii.	$r = \dfrac{1}{1 + \cos \theta}$
______iii	$r = \dfrac{2}{1 + \sin \theta}$	______ix.	$r = \dfrac{3}{1 - 2 \cos \theta}$
______iv.	$r = \dfrac{2}{1 - \sin \theta}$	______x.	$r = \dfrac{5}{1 + 3 \cos \theta}$
______v.	$r = \dfrac{4}{3 - \sin \theta}$	______xi.	$r = \dfrac{3}{1 + 2 \cos \theta}$
______vi	$r = \dfrac{4}{3 + \sin \theta}$	______xii.	$r = \dfrac{5}{1 - 3 \sin \theta}$

a. 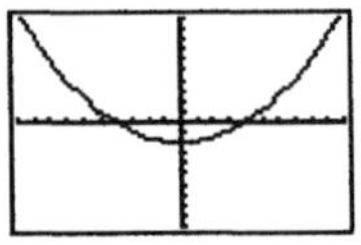
b. 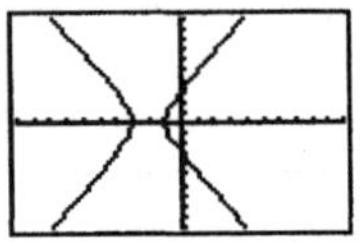
c. 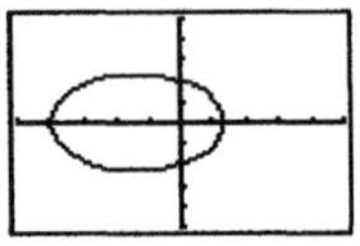
d. 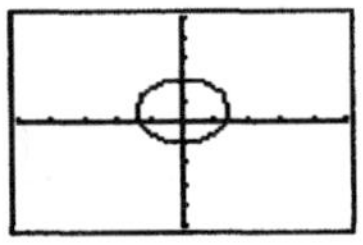
e. 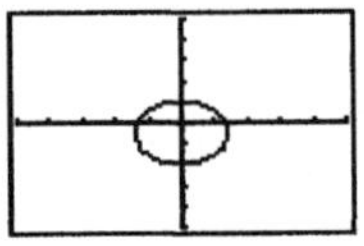
f. 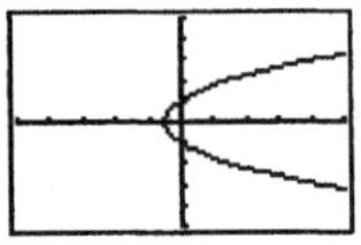

g. 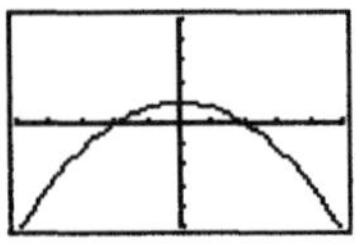
h. 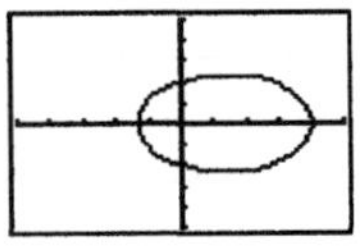
i. 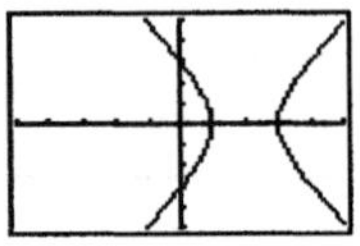
j. 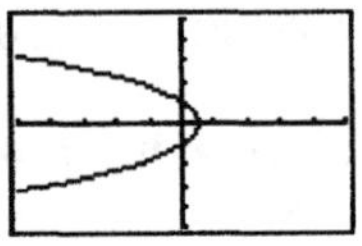
k. 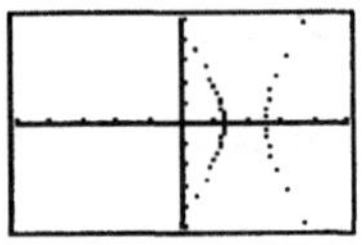
l. 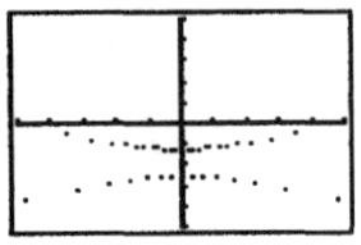

17. When necessary, write each of the equations in #17 in the form $r = \dfrac{ep}{1 \pm e \sin \theta}$ or $r = \dfrac{ep}{1 \pm e \cos \theta}$ and determine e's relationship (e>1, e=0,e<1) to the type of conic.

$e < 1$:______________________

$e = 1$:______________________

$e > 1$:______________________

Solutions: **1.** Answers may vary. **2**. Only the max and min values of x and y were the affected window values after zooming out.

6. $a < 0$ reflects the graph across the x-axis and across the y-axis, as $|a|$ increases so does the size of the spiral and as $|a|$ decreases the spiral decreases

8. $a < 0$ reflects the graph across the x-axis, as $|a|$ increases so does the size of the spiral and as $|a|$ decreases the spiral decreases

9 . If a is even then the number of leaves is $2a$, and if a is odd the number of leaves is a.

10. yes **12.** no, values in table do not correspond

13. If $a < b$, the graph increases in size as both a and b increase.
If $a = b$, the graph increases in size as both a and b increase.
If $a > b$, the graph increases in size as both a and b increase.

14. ZOOM IN until the window is approximately [-.04,.04] by [-.04,.04].

15. ZOOM OUT until the window is approximately [-10240, 10240] by [-10240,10240].

16. i. h **ii.** c **iii.** g **iv.** a **v.** d **vi.** e **vii.** f **viii.** j **ix.** b **x.** k **xi.** i **xii.** l

17. $e < 1$:ellipse, $e = 1$:parabola, $e > 1$: hyperbola

UNIT 27
PARAMETRIC GRAPHS OF CONICS

*Prerequisite: Unit #26

Parametric equations define a graph by expressing the variables *x* and *y* in terms of a third variable. The third variable is called the parameter. We will designate the third variable as *t* and express the variables *x* and *y* a functions of *t*: $f(t) = x$ and $g(t) =$.

Parametric equations allow us to represent as a function curves whose equations are not the graph of a function. To parameterize circles, ellipses and hyperbolas, we can define $f(t)$ and $g(t)$ in terms of trigonometric functions.

Recall that the six trigonometric functions defined by the right triangle with acute angle of measure *t* and radius of measure *r* as displayed at the right are as follows:

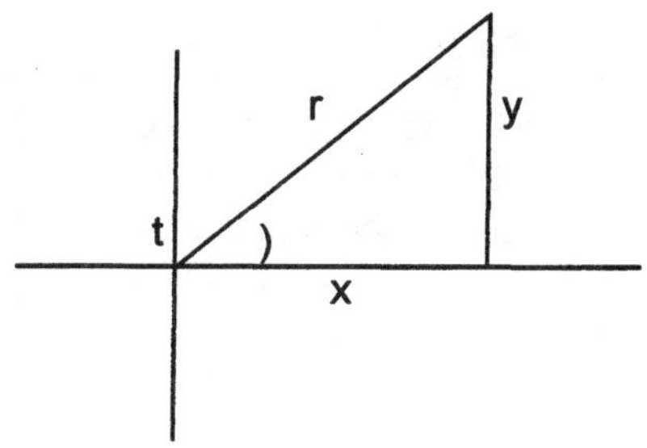

$\sin t = \dfrac{\text{opp}}{\text{hyp}} = \dfrac{y}{r}$ $\qquad \csc t = \dfrac{\text{hyp}}{\text{opp}} = \dfrac{r}{y}, y \neq 0$

$\cos t = \dfrac{\text{adj}}{\text{hyp}} = \dfrac{x}{r}$ $\qquad \sec t = \dfrac{\text{hyp}}{\text{adj}} = \dfrac{r}{x}, x \neq 0$

$\tan t = \dfrac{\text{opp}}{\text{adj}} = \dfrac{y}{x}, x \neq 0$ $\qquad \cot t = \dfrac{\text{adj}}{\text{opp}} = \dfrac{x}{y}, y \neq 0$

Note: Access the MODE screen of your calculator and ensure that all the options on the left are highlighted *except* Func. On that line, Par (or Parem, depending on your calculator) should be highlighted.

Circles

The equation of a circle centered at the origin with a radius of *r* can be parameterized by using the definition of the sine and cosine of the angle *t* given above:

$\cos t = \dfrac{x}{r}$ $\qquad \sin t = \dfrac{y}{r}$

$r \cdot \cos t = x$ $\qquad r \cdot \sin t = y$

and the trigonometric identity $\sin^2 t + \cos^2 t = 1$.

Example 1	**Graphically**	**Parameterization**
Find a parameterization of $x^2 + y^2 = 9$ and display the graph on the calculator.	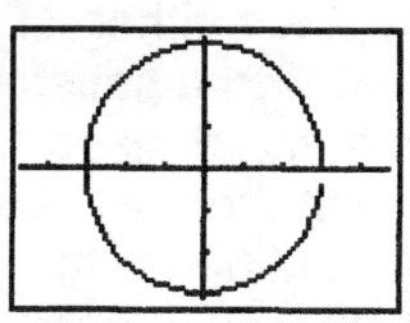 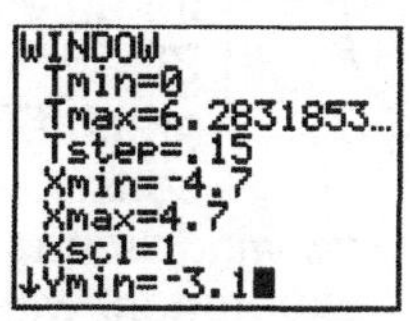 With the calculator in **Par**ametric MODE the parameterization is graphed in the ZDecimal viewing window.	The circle with equation $x^2 + y^2 = 3^2$ has a radius *r* equal to 3. The parameterization is $x = 3 \cos t$ and $y = 3 \sin t$ since $(3 \cos t)^2 + (3 \sin t)^2 =$ $9(\cos^2 t + \sin^2 t) =$ $9 \cdot 1 = 9 = 3^2$.

Parametric equations graphed with a graphing calculator must have the minimum and maximum values defined for the variable *t* as well as for the variables *x* and *y*. An incomplete graph can result from an inappropriate range for *t*. The increment for *t*, called the *t*-step, determines how many points are to be plotted when constructing the graph. Thus, a small *t*-step, between 0.05 and 0.15 in radian MODE, must be selected to produce a smooth graph.

Note: Because you are graphing a circle on a rectangular screen, you must be careful to select viewing windows that produce square results. On the TI-83/84 series, the ZDecimal and ZInteger windows produce square screens. Other windows may be squared up using the ZSquare feature under the ZOOM menu. TI-86 users will need to apply the ZSQR feature under the ZOOM menu to all screens given.

A circle with radius *r* and center at (h, k), $(x - h)^2 + (y - k)^2 = r^2$, has as its graph the parameterization

$$x = r\cos t + h \qquad y = r\sin t + k$$

Example 2	**Graphically**	**Parameterization**
Find a parameterization of $(x - 2)^2 + (y + 1)^2 = 9$ and graph on the calculator.	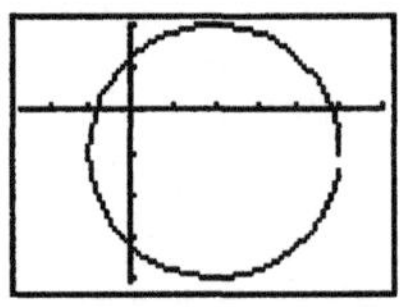 The graph is displayed in the window [-2.7,6] by [-4.1,2]. (The selected window is another version of the ZDecimal window. Note that the differences between the Xmax and Xmin and the Ymax and Ymin are the same as the differences between those same quantities in the ZDecimal window.	The equation has a radius of 3, $r = 3$, and its center is at (2, -1), $h = 2$ and $k = -1$. One parameterization would be $x = 3\cos t + 2$ and $y = 3\sin t - 1$.

The parameterization of a circle was based on the trigonometric identity $\sin^2 x + \cos^2 x = 1$. Trigonometric identities can be used to parameterize ellipses and hyperbolas.

Ellipses

The ellipse $\frac{x^2}{a^2} + \frac{y^2}{b^2} = 1$, with center at (0,0), can be parameterized by the equations

$x = a\cos t$ and $y = b\sin t$

since $\frac{(a\cos t)^2}{a^2} + \frac{(b\sin t)^2}{b^2} = \frac{a^2\cos^2 t}{a^2} + \frac{b^2\sin^2 t}{b^2} = \cos^2 t + \sin^2 t = 1$.

Example 3	**Graphically**	**Parameterization**
Find a parameterization of the ellipse with the equation $\frac{x^2}{4} + \frac{y^2}{9} = 1$ and graph on the calculator.	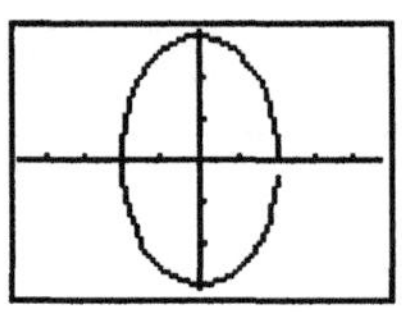 A square window is not essential for an elllipse, thus either ZTrig or ZDecimal produces a satisfactory graph (ZDecimal was used for consistency with the previous examples.) The window has been defined with $0 \le T \le 2\pi$. Experiment with *T*step values between .05 and 0.15. Textbooks will sometimes suggest a *T*step of $\frac{\pi}{24}$.	From the equation, $a = 2$ and $b = 3$. Thus, one parameterization would be $x = 2\cos t$, $y = 3\sin t$.

An ellipse $\frac{(x-h)^2}{a^2}+\frac{(y-k)^2}{b^2}=1$, with center at (h, k), can be parameterized to

$x = a\cos t + h$ and $y = b\sin t + k$.

Example	Graphically	Parameterization
Find a parameterization of $\frac{(x-2)^2}{9}+\frac{(y-3)^2}{16}=1$ and graph on the calculator.	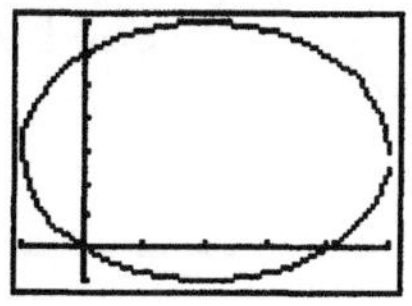 Neither the ZDecimal or ZTrig window are satisfactory. The window displayed is [-1, 5] by [-1, 7].	This ellipse has as its center (2,3). Thus, $h = 2$, $k = 3$, $a = 3$, and $b = 4$, yielding a parameterization of $x = 3\cos t + 2$ and $y = 4\sin t + 3$.

Hyperbola

A hyperbola $\frac{x^2}{a^2}-\frac{y^2}{b^2}=1$ with center at (0,0) can be parameterized by

$$x = a\sec t = \frac{a}{\cos t} \quad \text{and} \quad y = b\tan t,\ 0 \le t \le 2\pi$$

using the trigonometric identity $\tan^2 t + 1 = \sec^2 t$. Similarly, a hyperbola with center at (h, k), $\frac{(x-h)^2}{a^2}-\frac{(y-k)^2}{b^2}=1$, can be parameterized by

$$x = a\sec t + h = \frac{a}{\cos t}+h \quad \text{and} \quad y = b\tan t + k\ (0 \le t \le 2\pi).$$

Example	Graphically	Parameterization
Find a parameterization and calculator graph of $\frac{(x+1)^2}{9}-\frac{(y-2)^2}{16}=1$.	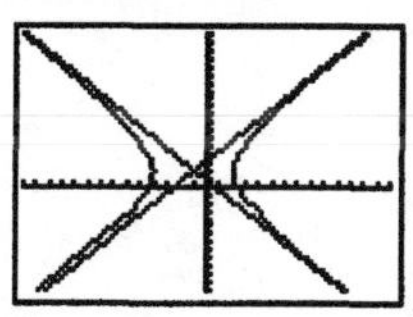 Graphed in the window [-14,14] by [-13, 19] we appear to also have the asymptotes created by the fundamental rectangle that is constructed when we graph by hand. Remember, at $\frac{\pi}{2}$ and $\frac{3\pi}{2}$ these functions are not defined. Watch carefully as points are plotted on the screen. The calculator is actually connecting two points on either side of the asymptote. If the graphing mode is changed from connected to dot these "lines", which are indeed not part of the graph, will not appear. 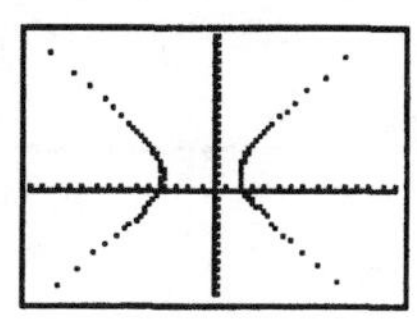	This hyperbola has its center at (-1,2). Thus, $h = -1$, $k = 2$, $a = 3$ and $b = 4$, yielding $x = \frac{3}{\cos t} - 1$ and $y = 4\tan t + 2$ as one parameterization.

An hyperbola $\frac{y^2}{a^2} - \frac{x^2}{b^2} = 1$ with center at (0,0) can be parameterized by

$x = b\ \tan t$ and $y = \frac{a}{\cos t}$, $0 \le t \le 2\pi$.

Similarly, an hyperbola with center at (h, k), $\frac{(y-k)^2}{a^2} - \frac{(x-h)^2}{b^2} = 1$, can be parameterized by

$x = b \tan t + h$ and $y = \frac{a}{\cos t} + k$ $(0 \le t \le 2\pi)$.

Example	**Graphically**	**Parameterization**
Find a parameterization and calculator graph of $\frac{(y+1)^2}{9} - \frac{(x-2)^2}{16} = 1$.	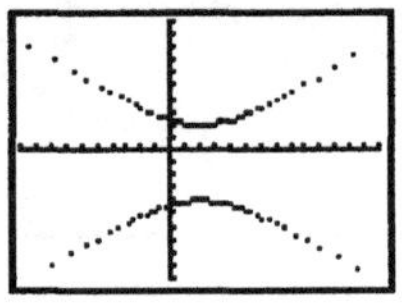 The graph is displayed in the window [-10,14] by [-10, 10] in dot mode.	This hyperbola has its center at (2,-1). Thus, $h = 2$, $k = -1$, $a = 3$ and $b = 4$, yielding $x = 4 \tan t + 2$ and $y = \frac{3}{\cos t} - 1$ as one parameterization.

EXERCISE SET

Directions: Determine a parameterization where $0 \le T \le 2\pi$ for each equation below and sketch the graph. Record window values for each graph.

1. $x^2 + y^2 = 36$

   x = ________________ y = ________________

   [____, ____] by [____, ____]

2. $x^2 + (y + 3)^2 = 25$

   x = ________________ y = ________________

   [____, ____] by [____, ____]

3. $(x - 2)^2 + (y - 5)^2 = 4$

   x = ________________ y = ________________

   [____, ____] by [____, ____]

4. $\frac{x^2}{4}+\frac{y^2}{9}=1$

$x =$ ________________ $y =$ ________________

[____ , ____] by [____ , ____]

5. $\frac{(x-4)^2}{25}+\frac{(y-1)^2}{36}=1$

$x =$ ________________ $y =$ ________________

[____ , ____] by [____ , ____]

6. $\frac{(x+3)^2}{16}+\frac{(y-2)^2}{25}=1$

$x =$ ________________ $y =$ ________________

[____ , ____] by [____ , ____]

7. $\frac{x^2}{9}-\frac{y^2}{4}=1$

$x =$ ________________ $y =$ ________________

[____ , ____] by [____ , ____]

8. $\frac{y^2}{4}-\frac{x^2}{9}=1$

$x =$ ________________ $y =$ ________________

[____ , ____] by [____ , ____]

9. $\frac{(x-2)^2}{49}-\frac{y^2}{16}=1$

$x =$ ________________ $y =$ ________________

[____ , ____] by [____ , ____]

10. $\frac{(y-3)^2}{4}-\frac{(x+1)^2}{9}=1$

$x =$ ________________ $y =$ ________________

[____ , ____] by [____ , ____]

Solutions: **1.** $x = 6 \cos t$, $y = 6 \sin t$; [-10.6, 10.6] by [-7, 7]

2. $x = 5 \cos t$, $y = 5 \sin t - 3$; [-7.6, 7.6] by [-8, 2]

3. $x = 2 \cos t + 2$, $y = 2 \sin t + 5$; [-3.3, 7.3] by [0, 7]

4. $x = 2 \cos t$, $y = 3 \sin t$; [-4.7, 4.7] by [-3.1, 3.1]

5. $x = 5 \cos t + 4$, $y = 6 \sin t + 1$; [-10, 10] by [-10, 10]

6. $x = 4 \cos t - 3$, $5 \sin t + 2$; [-10, 10] by [-10, 10]

7. $x = \dfrac{3}{\cos t}$, $y = 4 \tan t$; [-4.7, 4.7] by [-3.1, 3.1]

8. $x = 2 \tan t$, $y = \dfrac{3}{\cos t}$; [-10, 10] by [-10, 10]

9. $x = \dfrac{7}{\cos t} + 2$, $y = 4 \tan t$; [-10, 15] by [-10, 10]

10. $x = 2 \tan t + 3$, $y = \dfrac{-3}{\cos t} - 1$; [-10, 15] by [-10, 10]

UNIT 28
ROOTS OF COMPLEX NUMBERS

***Prerequisite:** Unit #27.

This unit examines nth roots of complex numbers in trigonometric form. If a + bi is a complex number, then the trigonometric form of that number is r (cos θ + i sin θ) where $r = |a+bi| = \sqrt{a^2+b^2}$. De Moivre's Theorem, used for computing integral powers of complex numbers, states:

if n is any real number, then $[r(\cos\theta + i \sin\theta)]^n = r^n(\cos r\theta + i \sin r\theta)$.

An extension of De Moivre's Theorem makes it possible to compute the n nth roots of a complex number in trigonometric form:

$$\sqrt[n]{r}\left(\cos\frac{\theta+k\cdot 360^\circ}{n}+i\sin\frac{\theta+k\cdot 360^\circ}{n}\right) \text{ where } k = 0,1,2,3,\ldots,n-1.$$

Example 1: Find the five fifth roots of $-\sqrt{2}+\sqrt{2}i$.

Solution: Express $-\sqrt{2}+\sqrt{2}i$ in trigonometric form: 2(cos 135° + i sin 135°), and apply the extension of De Moivre's Theorem.

$$\sqrt[5]{2}\left[\cos\frac{135+k\cdot 360^\circ}{5}+i\sin\frac{135+k\cdot 360^\circ}{5}\right]$$

Each of the five roots r_1, r_2, r_3, r_4, r_5 is determined by letting k = 0,1,2,3,4. Thus

$\sqrt[5]{2}\,(\cos 27^\circ + i \sin 27^\circ) \approx 1.0235 + .5215i$

$\sqrt[5]{2}\,(\cos 99^\circ + i \sin 99^\circ) \approx -.1797 + 1.1346i$

$\sqrt[5]{2}\,(\cos 171^\circ + i \sin 171^\circ) \approx -1.1346 + .1797i$

$\sqrt[5]{2}\,(\cos 243^\circ + i \sin 243^\circ) \approx -.5215 - 1.0235i$

$\sqrt[5]{2}\,(\cos 315^\circ + i \sin 315^\circ) \approx .8122 - .8122i$

Observe that each of the roots r_1 through r_5 has an absolute value of approximately 1.15. Thus, these roots are located on a circle with center at the origin and radius of 1.15.

(Note also that $\sqrt[5]{2} \approx 1.15$.)

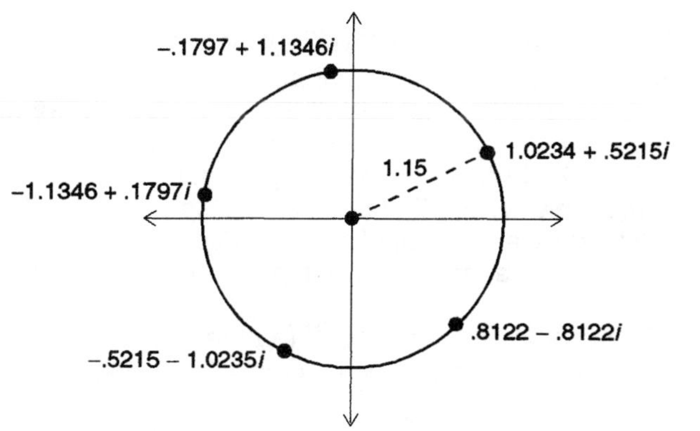

◆

In general, the nth roots of a complex number z are located at equal intervals on a circle of radius $\sqrt[n]{r}$ where $z = r(\cos\theta + i \sin\theta)$.

These roots can be displayed both graphically and in table form when the calculator is in parametric MODE.

Table

Set the calculator to both **Degree** and **Parametric** MODE.

To display the roots of complex numbers using a circle whose radius is $\sqrt[n]{r}$, recall that the parametric equations $x = r\cos t$ and $y = r\sin t$ define the graph of a circle $x^2 + y^2 = r^2$. To determine the *n nth* roots of a complex number, let X1T = $\sqrt[n]{r}$ cos T and Y1t = $\sqrt[n]{r}$ sin T.

Press **[Y=]** and enter $\sqrt[5]{2}$ cos T after X1T= and $\sqrt[5]{2}$ sin T after Y1T= to determine the five fifth roots of $-\sqrt{2}+\sqrt{2}i$ that were previously computed "by hand."

To view these roots in the TABLE, set **TblStart** = 27 and **ΔTbl** = 360/5. In general, set **TblStart** =θ/n and **ΔTbl** = 360/n.

NOTE: The angle θ must be the <u>actual</u> angle, not the <u>reference</u> angle.

Display the TABLE as shown at the right. The X1T column provides the *a* values in the complex number *a + bi* and the Y1T column provides the *b* values.

T	X1T	Y1T
27	1.0235	.5215
99	-.1797	1.1346
171	-1.135	.1797
243	-.5215	-1.023
315	.81225	-.8123
387	1.0235	.5215
459	-.1797	1.1346

T=27

Complex Roots:
1.0235 + .5215 i
-.1797 + 1.1346 i
-1.135 - .1797 i
-.5215 - .1.023 i
.81225 - .8123 i

Graph

To display the graph of the circle with radius $\sqrt[5]{2}$ that contains these roots, the WINDOW should be set with Tmin=0, Tmax=360 and Tstep=360/5 (in general, 360/n). Initially set xMin/xMax and yMin/yMax to $\pm\sqrt[5]{2}$ (in general, $\pm\sqrt[n]{r}$) accordingly.

Press **[GRAPH]** to display the graph at the right. The graph is not displayed as a circle because insufficient points are graphed when the Tstep is too large.

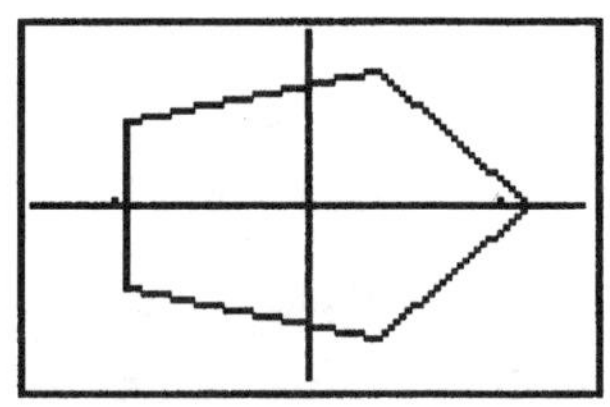

To adjust the WINDOW settings so that a circle of radius $\sqrt[5]{2}$ is displayed, return to Example1 and observe that the initial angle is 27°, with subsequent angles occurring at 72° intervals. If the Tstep is a divisor of both 27 <u>and</u> 72, then the initial angle of 27° will appear on the circle as well as all angles of the form 27 + 72k. In general, the Tstep must be a divisor of θ and 360/n. Because 9 is the largest divisor of 27 and 72, the Tstep was set to 9 for the display at the right and **ZSquare** was then applied to produce a true circle.
TRACE to verify the remaining four fifth roots.

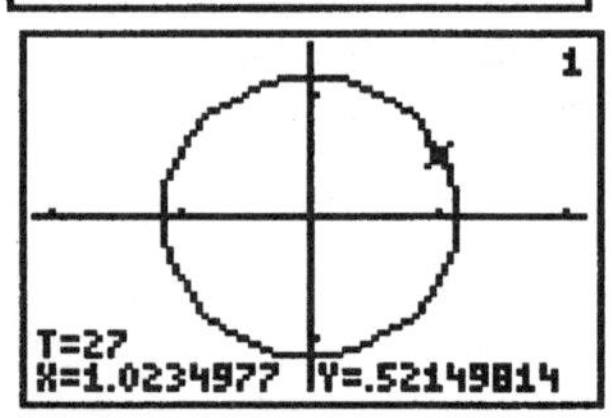

Example 2: Use the TABLE feature to determine the five fifth roots of 1 (1+0i), also called the five fifth roots of unity. Fill in the values on the TABLE and record the roots in a + bi form.

Solution: $\sqrt[n]{r}=\sqrt[5]{1}$. Begin by letting X1T = $\sqrt[5]{1}$ cos T and Y1T = $\sqrt[5]{1}$ sin T. Set the table to start at 0 and increment by 72 (360/5).

The roots, as displayed in the table, are:

T	X1T	Y1T
0	1	0
72	.30902	.95106
144	-.809	.58779
216	-.809	-.5878
288	.30902	-.9511
360	1	0
432	.30902	.95106

T=0

These roots are 1, 0.3090 + .9511i, -0.8090 + 0.5878i, -0.8090 - 0.5878i, and 0.3090 - 0.9511i.

To graphically display the five fifth roots of 1 on a unit circle, the Tstep will need to be a divisor of 360/5 = 72 and θ = 0. Since 0 is divisible by all real numbers except itself, any divisor of 72 is an acceptable Tstep value.

The divisors of 72 are 1,2,3,4,6,8,9,12,18,24,36, and 72. A Tstep of 72 does not produce a circle, however, the roots are clearly apparent. A Tstep value less than or equal to 24 will produce a unit circle, tracing will verify the roots displayed in the table.

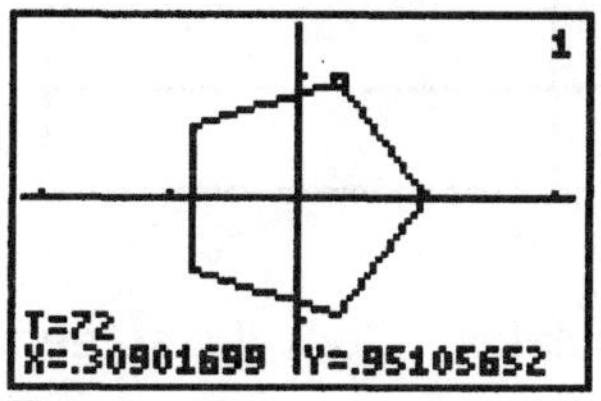

Tstep = 72

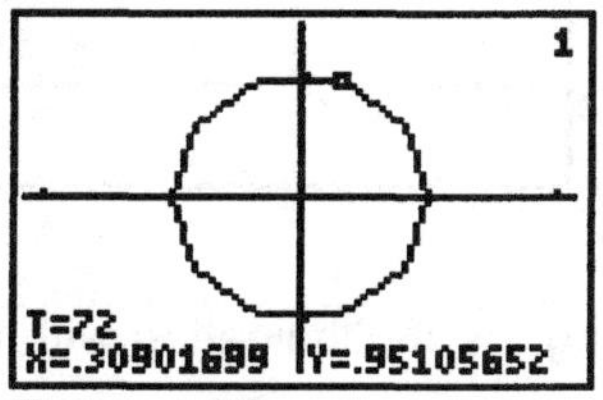

Tstep = 36

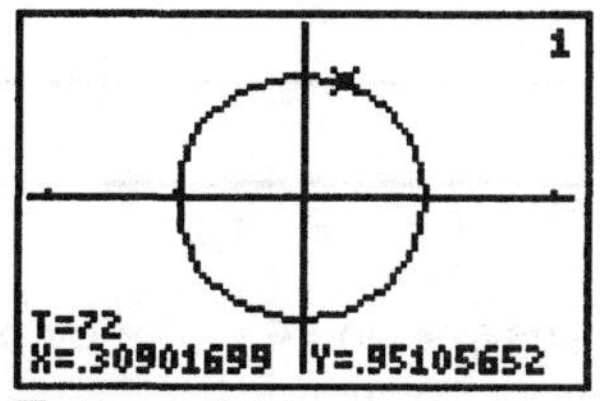

Tstep = 24

*NOTE:*Another form of the original problem would be stated $x^5 = 1$.

◆

EXERCISE SET

Directions: Display the roots in TABLE form, completing the displayed table, and then record the roots in a + bi form.

1. Find the four fourth roots of $-8 - 8\sqrt{3}i$, i.e. solve $x^4 = -8 - 8\sqrt{3}i$.

ROOTS ARE:

____

____

____

____

T	X1T	Y1T

2. Find the four fourth roots of $-8 + 8\sqrt{3}i$, i.e. solve $x^4 = -8 + 8\sqrt{3}i$.

ROOTS ARE:

____

____

____

____

T	X1T	Y1T

3. Compare the solutions to $x^4 = -8 - 8\sqrt{3}i$ i and $x^4 = -8 + 8\sqrt{3}i$. Based on your observations, if the solutions to $x^5 = -\sqrt{2} + \sqrt{2}i$ are

1.0235 + .5215i   -1.1346 + .1797i   .8122 - .8122i
-.1797 + 1.1346i   -.5215 - 1.0235i,

can you write the solutions to $x^5 = -\sqrt{2} - \sqrt{2}i$?

____________________   ____________________   ____________________

____________________   ____________________

Verify these roots with the calculator.

4. Find the two square roots of -4i.

ROOTS ARE:

____

____

T	X1T	Y1T

5. Verify the answers in #4 either by performing the multiplication with the calculator . Copy the screen display of your calculator.

6. Find the four fourth roots of unity.

ROOTS ARE:

____

____

____

____

T	X1T	Y1T

7. Find the eight eighth roots of unity.

ROOTS ARE:

____

____

____

____

____

____

____

____

T	X1T	Y1T

✍8. Based on the results of #6 & 7, can a partial list of roots be determined for the twelve twelfth roots of unity or even the 4n 4nth roots of unity? In general, if you know the roots of unity for index "k", is it possible to create an accurate partial list of roots of unity for index kn?

✍9. Based on the results of #1,2 & 3, is it possible to formulate an algorithm for computing like roots of complex conjugates? Since these three problems only addressed fourth roots, be sure to explore other roots before drawing any conclusions.

Solutions:

1. $n=4$, $r=16$, $\sqrt[n]{r}=2$, TblStart=60, ΔTbl=90
   1+1.7321i, -1.732 +i, -1 - 1.732i, 1.7321 - i

2. $n=4$, $r=16$, $\sqrt[n]{r}=2$, TblStart=30, ΔTbl=90
   1.7321 + i, -1 + 1.7321i, -1.732 - i, 1 - 1.732i

3. Solutions are complex conjugates of those listed.

4. $n=2$, $r=4$, $\sqrt[n]{r}=2$, TblStart=135, ΔTbl=180
   -1.4142 + 1.4142 i, 1.414 - 1.414i

5. Remember as you verify: you are computing with approximations and thus your result will be "close" but not exact.

6. $n=4$, $r=1$, $\sqrt[n]{r}=1$, TblStart=0, ΔTbl=90
   1 +0i, 0 + i, -1 + 0i, 0 - i

7. $n=8$, $r=1$, $\sqrt[n]{r}=1$, TblStart=0, ΔTbl = 45
   1 + 0i, .7071 + .7071i, 0 + i, -.7071 + .7071i, -1 + 0i, -.7071 - 7071i, 0 - i,
   .7071 - .7071i

8. Answers may vary.

9. Answers may vary.

INDEX